GREAT WISDOM IN SMALL SPACE

小空间大智慧

THE GREEN DESIGN OF SMALL-SIZE ARCHITECTURE

小型绿色建筑设计实践

ThinkArchit工作室 主编

華中科技大學出版社
http://www.hustpas.com

小空间 大智慧

GREAT WISDOM IN SMALL SPACE

The Green Design of Small-size Architecture

小型绿色建筑设计实践

PREFACE 前 言

本书在全球范围内精选了50多个优秀的小面积建筑设计作品，其中包括住宅建筑和公共设施建筑。这些建筑设计理念独特，不仅是将有限空间合理利用的典范，更是绿色节能建筑设计的代表。建筑师通过借鉴这些实际案例，能更好地掌握建筑设计的原则，并体会建筑空间组合的技巧。那些节能、环保的设计方法以及节能材料的应用形式，为建筑师探求绿色建筑设计方法提供了参考和依据。本书展示的每个作品都配有详细的设计说明、平面图、立面图、设计构思示意图、建成照片等，真实展现了每个作品的设计理念和设计方法。

本书由ThinkArchit工作室策划编辑，在编写过程中，得到了众多建筑设计事务所和建筑设计师的大力支持，在此表示感谢。

由于编者水平有限，错误和不足之处在所难免，恳请各位读者指正。

The book covers more than 50 outstanding small-size architectural works in the world, including residential buildings and public facilities. The architectural design is unique. It is not only a model of optimum utilization of limited space but also representative of the green energy-saving building design. By referring to these cases, architects will have a better understanding of architectural design principles and appreciate the idea of how to combine space of the building. Those energy-saving and environment-friendly design and application of energy-saving materials provide a reference and basis for architects to explore green design. Each work in the book is displayed with detailed design specification, floor plans, facades, diagrams and photos, which show the design concept and method of each work.

The book is edited by ThinkArchit Studio. In the preparation process, the editors received strong support from a number of architectural firms. Many thanks for their help. As the editor is limited, errors and deficiencies are inevitable. Please let us know of any errors or deficiencies you find, as well as your suggestions.

CONTENTS 目 录

Residence

Public Facility

小空间 大智慧

GREAT WISDOM IN SMALL SPACE

The Green Design of Small-size Architecture

小型绿色建筑设计实践

RESIDENCE 住宅

Candela House

Zigzag Arquitectura

■ Architecture Design/建筑设计: Zigzag Arquitectura ■ Project Architect/项目建筑师: David Casino, Bernardo Angelini ■ Location/地点: Principado de Asturias, Spain ■ Area/面积: 297.98 m^2
■ Completion/建成时间: 2009 ■ Photograph/摄影: B. Angelini , D. Casino

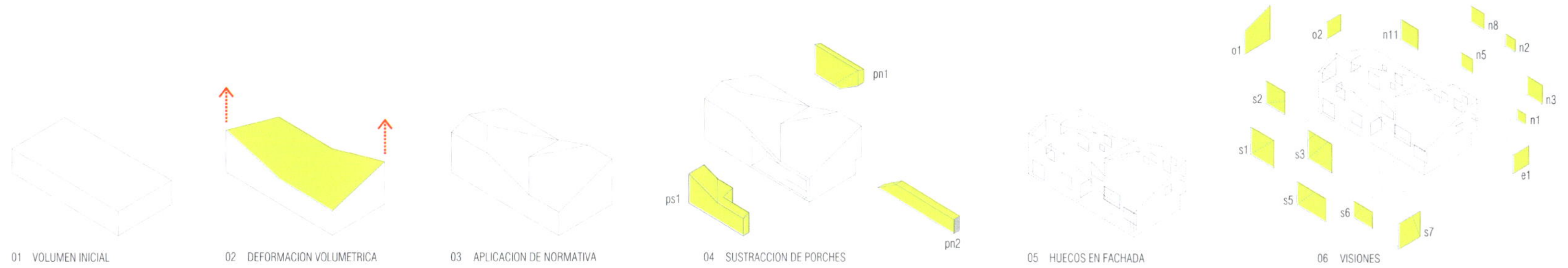

A family meeting place. The starting point was the need for a house large enough for the variable and regular meeting of a family composed of 6 subfamilies, with a total of 10 adults and 8 children. The idea was to create a family shelter, a place to meet and escape from urban life, with a low budget and the scarce construction resources available in a relatively isolated location.

The place / an eyed mountain. The plot is located in an area surrounded by mountains, Aballe Dego, close to the River Sella. The place suggested the idea of a mountain, hollowed inside, with eyes strategically arranged to capture the surrounding landscape, introducing it into the rooms of the house and avoiding direct views into the homes of the nearest neighbours.

Volume and outer skin. The volume interacts with the surrounding mountains, longing to become one of them. The folded roof is also a product of the enforcement of the local regulations and the hipped roof indicated by them. The result is a hard shell, earthy and perforated, allowing the entry of the landscape in the form of pictures which change each time you return to the house. The perforated skin protects its inhabitants from the outside, while allowing them to interact physically and visually with nature, with the seasons and with the weather.

Programme. The programme is organized according to the dominant orientation of the plot. The night programme, bedrooms and bathrooms, is located in the northern area, leaving the southern orientation to the meeting spaces and the double height space: living room - dining room - kitchen. The porch serves as a link between the day spaces of the ground floor, and at the same time is used to mark the difference with the kitchen-dining room-living room programme.

Material and construction system. The construction system is adapted to the resources available in the area: reinforced concrete structure as well as double brick isolated walls.The facades of the house are made with a dark grey rendering coat and planks of iroko wood. The roof is built with flat concrete tiles, adapting to the material regulations of the area.

家庭会议室 这个设计的出发点，是考虑到这个大家庭由6个小家庭组成，总共10个大人和8个孩子。为了满足一家人临时或是定期家庭会议的需要，这座房子要有足够大的空间。设计师的理念是建造一个家庭的庇护之地，一家人能聚在一起， 在这里躲避城市的嘈杂。然而，限制因素在于，该预算成本低，而且项目位于相对偏远的地带，建造资源也相对匮乏。

选址与山景 该址位于群山Aballe Dego之中，临近Sella河。这样的地理位置启发我们想到了一个设计方案：在山谷中间的凹地中建房，这样在室内的每一个房间都能领略到四周的风景，不仅如此，还能避免直视周围的邻居。

体积与外表 建筑与周围的山峦交相辉映，浑然一体。折叠的屋顶和有斜脊的屋顶也是按照当地的细则实施而建，因此建筑外部非常坚实。每当房子的主人回到家的时候，都能看到入口处如画般的景观。穿孔式的建筑外表能够保护居民的隐私，同时也使该住户与外界大自然息息相通。

方案 该方案主要是根据朝东的地势而安排的。晚间活动区、卧室和浴室都安排在北边的区域，将日常活动区如客厅、餐厅、厨房安排在南边，而且南边的房高是北边的两倍。门廊在这里起到了双重作用，不仅是连接一层日常生活区域的纽带，也是区别于厨房、餐厅、客厅的分水岭。

材料和建筑体系 在该项目中，与当地资源相适应的建筑体系有钢筋混凝土结构和双倍的石砖隔离墙，建筑表面的颜色则以深灰色为主。建筑屋顶使用的是平滑的混凝土瓦片，与该地规定使用的建筑材料相匹配。

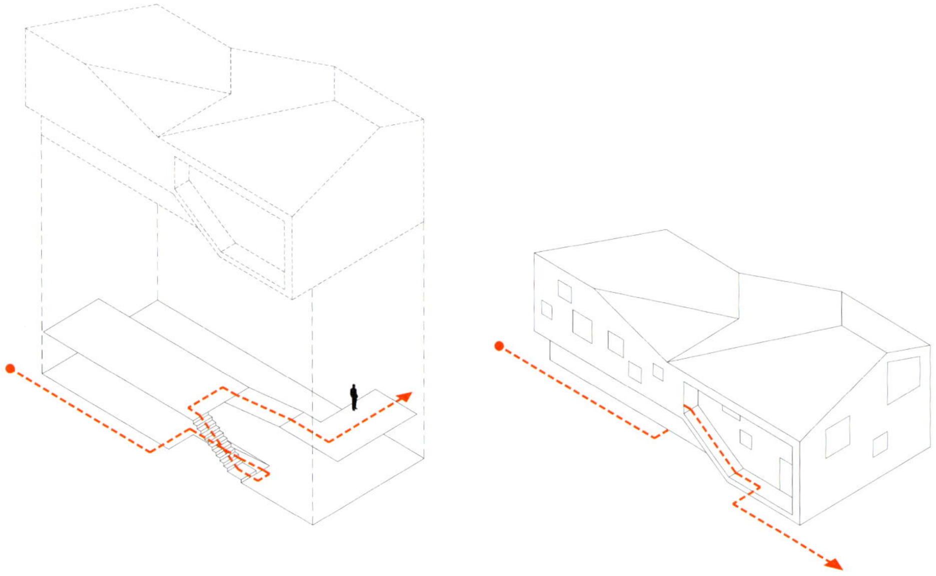

Diagram/示意图

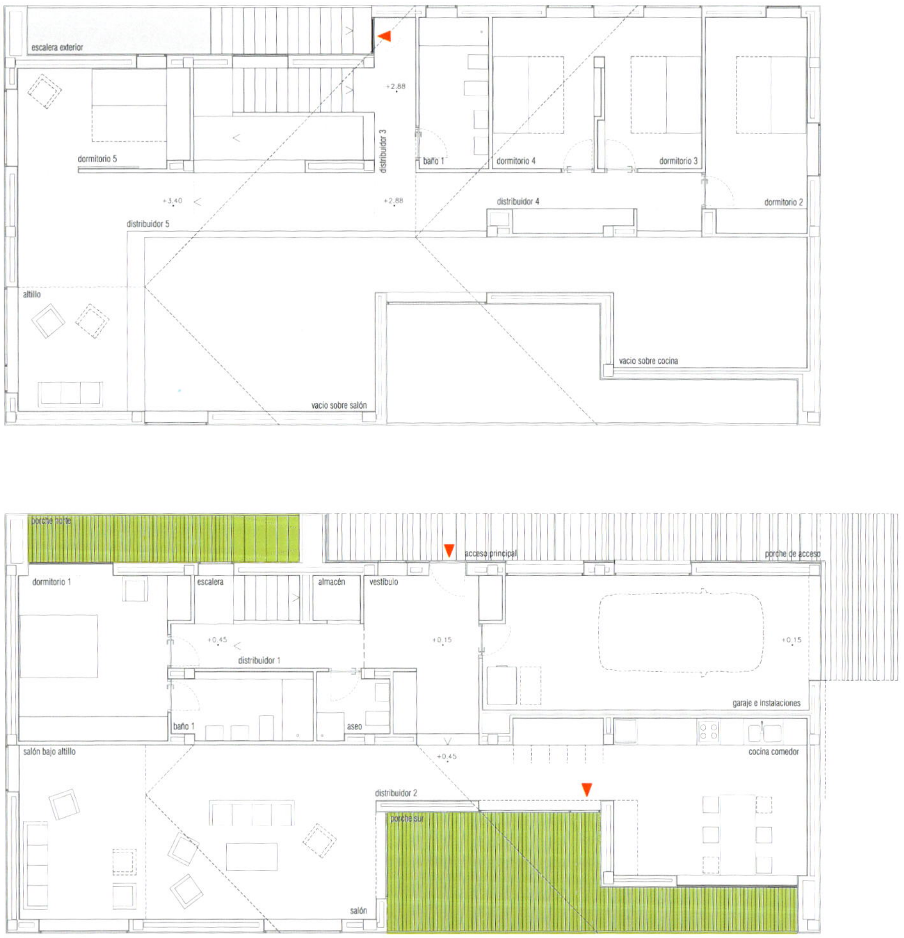

Plan/平面图

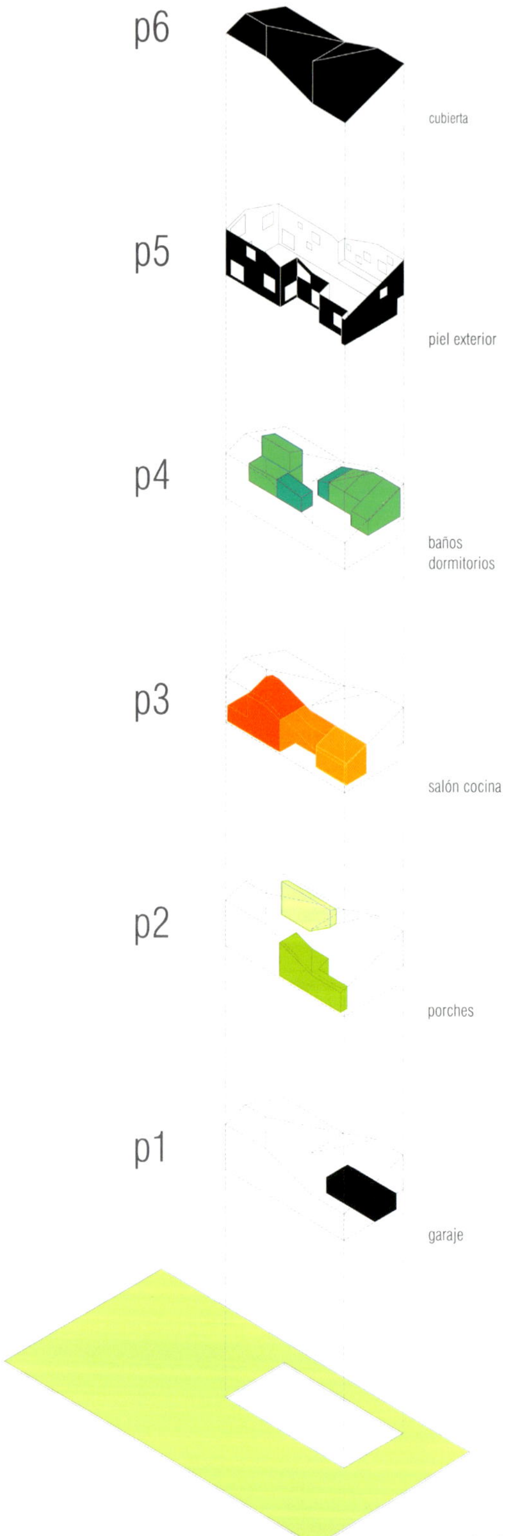

Diagram/示意图

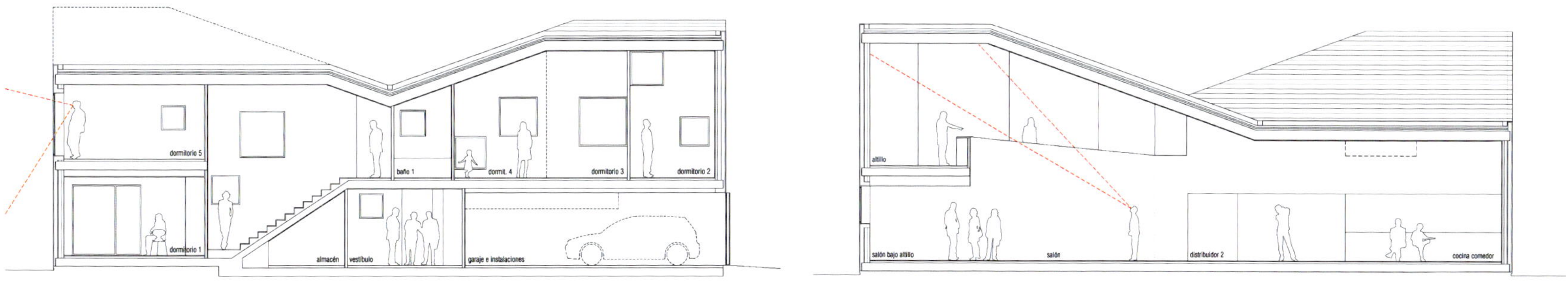

Section/剖面图

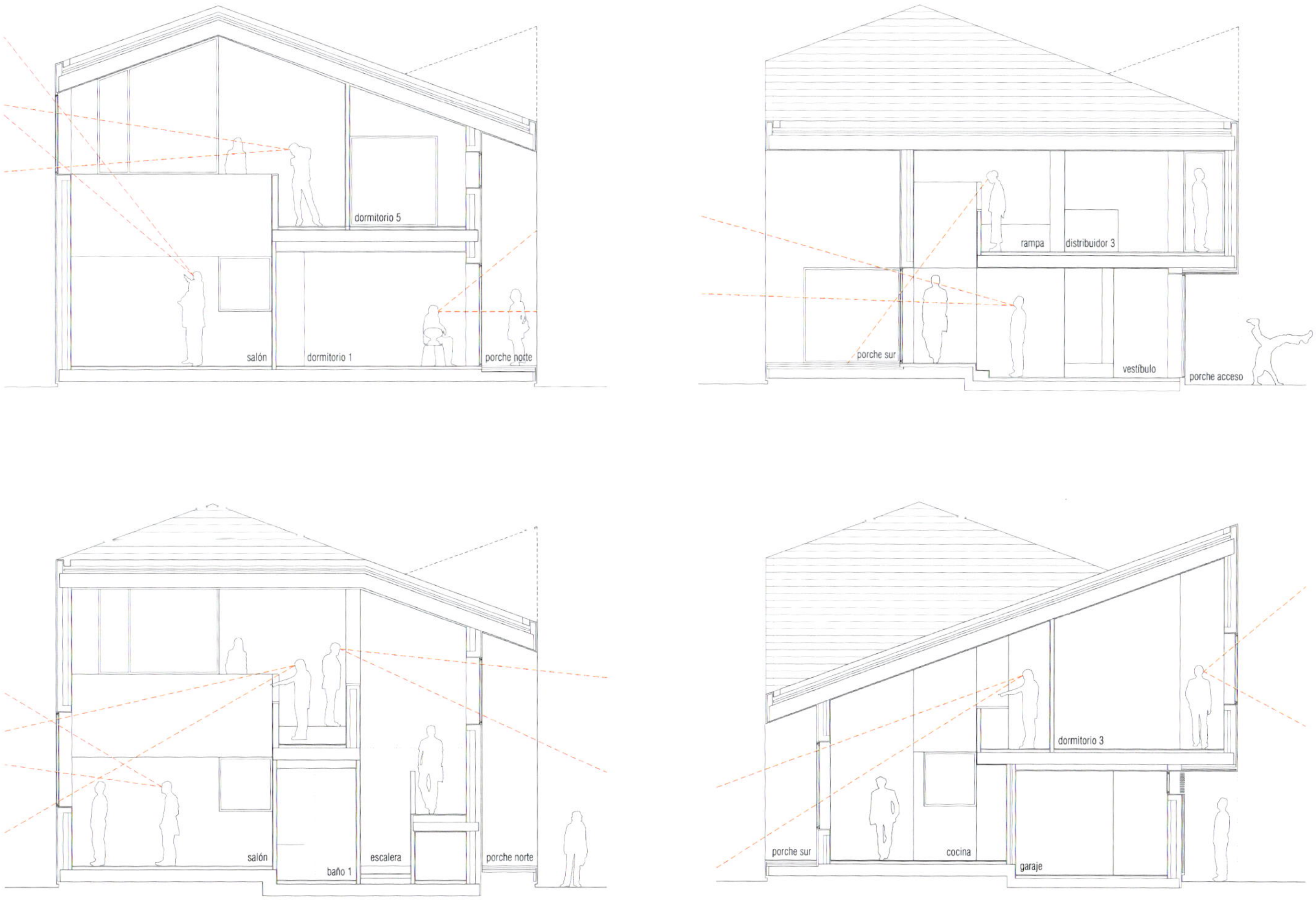

Section/剖面图

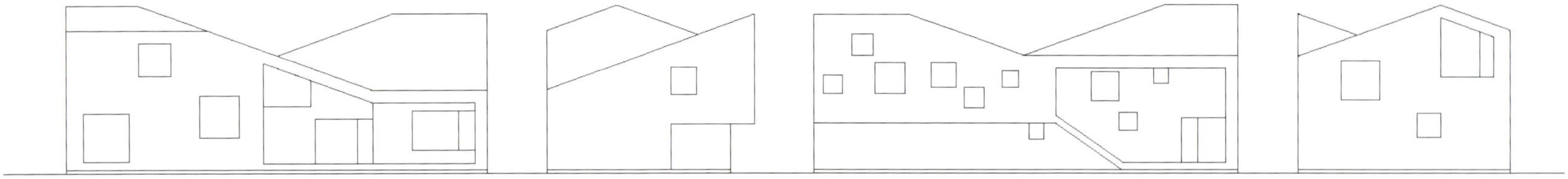

Facade/立面图

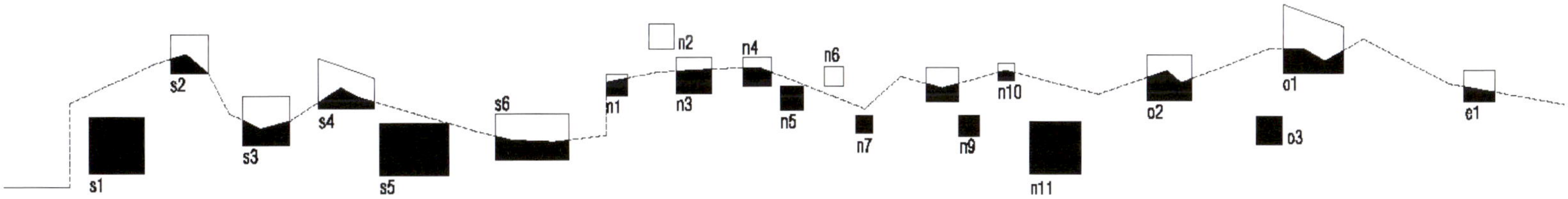
s1
s2
s3
s4
s5
s6
n1
n2
n3
n4
n5
n6
n7
n9
n10
n11
o1
o2
o3
e1

Dupli.Casa - House

J. MAYER H. Architects

■ Architecture Design/建筑设计: J. MAYER H. Architects ■ Design Team/设计团队: Juergen Mayer H., Georg Schmidthals, Thorsten Blatter, Simon Takasaki, Andre Santer, Sebastian Finckh
■ Location/地点: Germany ■ Area/面积: site 6,900 m^2/building 569 m^2/total floor 1,190 m^2 ■ Completion/建成时间: 2008 ■ Photograph/摄影: David Franck

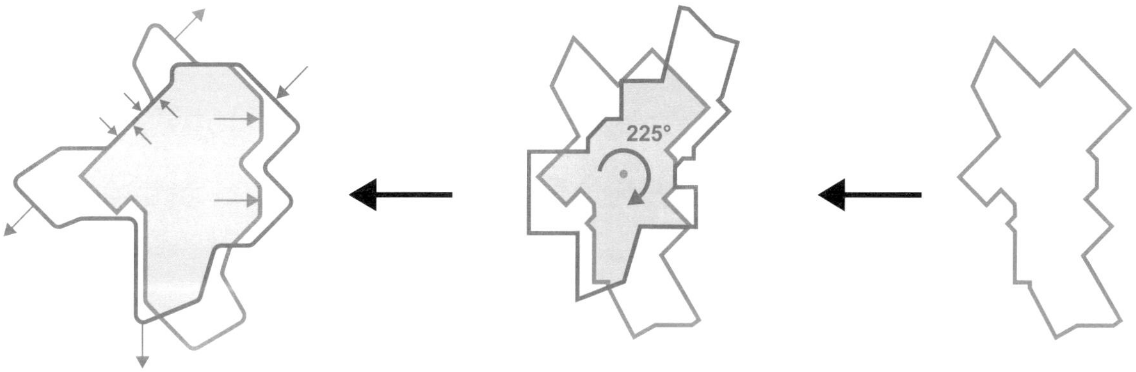

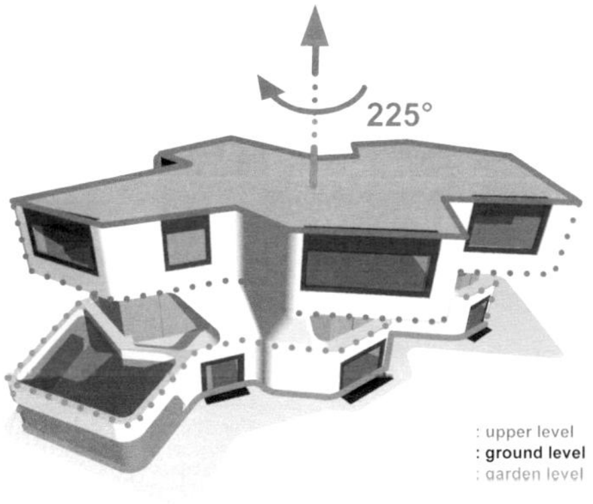

The geometry of the building is based on the footprint of the house that previously was located on the site. Originally built in 1984 and with many extensions and modifications since then, the new building echoes the "family archaeology" by duplication and rotation. Lifted up, it creates a semi-public space on ground level between two layers of discretion. The skin of the villa performs a sophisticated connection between inside and outside and offers spectacular views onto the old town of Marbach and the German national literature archive on the other side of the Neckar valley.

这个几何形建筑是在原有房屋的遗址上重建的。原始建筑建于1984年，之后，这个建筑就不断地被扩展和翻修，新建筑物复制和旋转形式的应用，体现了对原有建筑的尊重与追忆。这个别墅的设计别具匠心地将建筑物的一层抬高，在地面与一层之间开辟了一个半开放的空间。这栋别墅的外表成功地将室内和室外风格统一起来，成为马尔巴赫这个古老的小镇和对面内卡河山谷上的德国国家文献档案馆的一个壮观的景致。

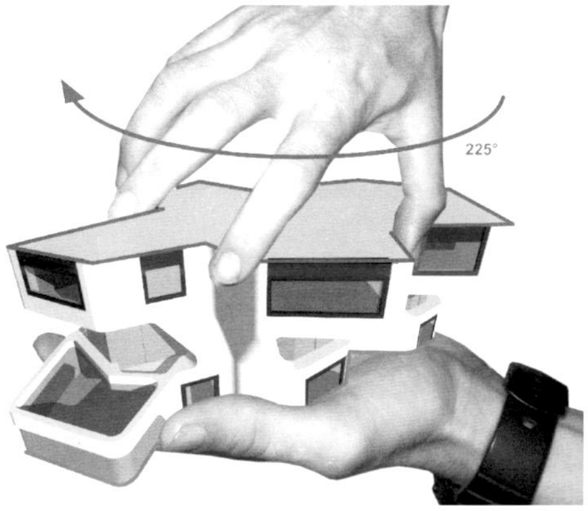

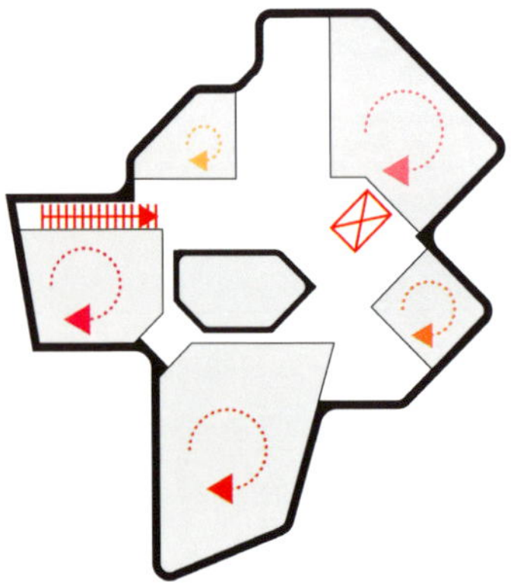

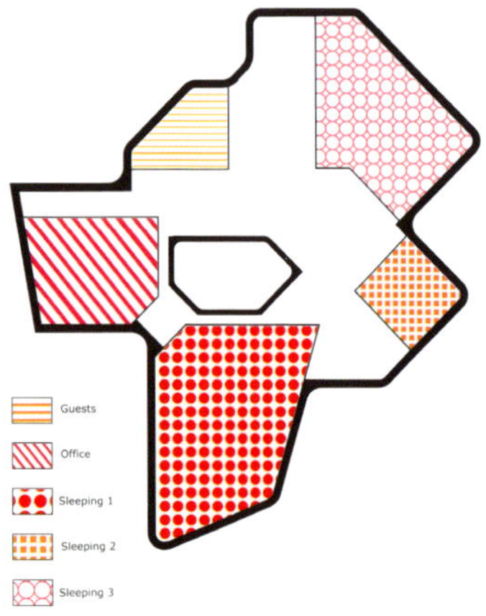

Diagram/示意图

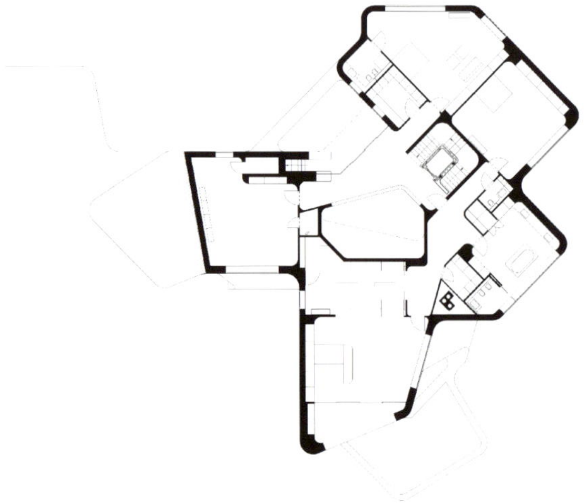

Second Floor Plan/ 二层平面图

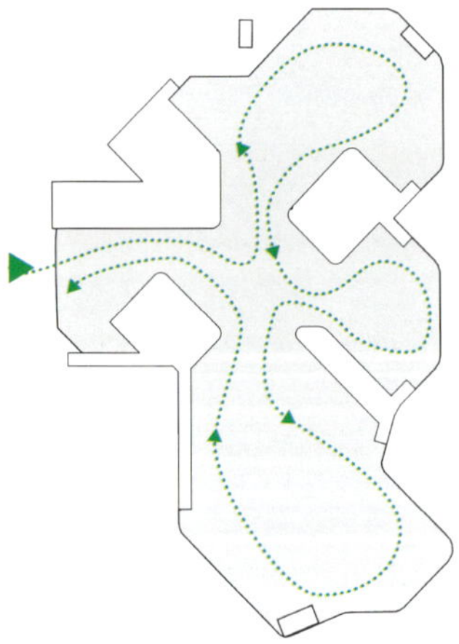

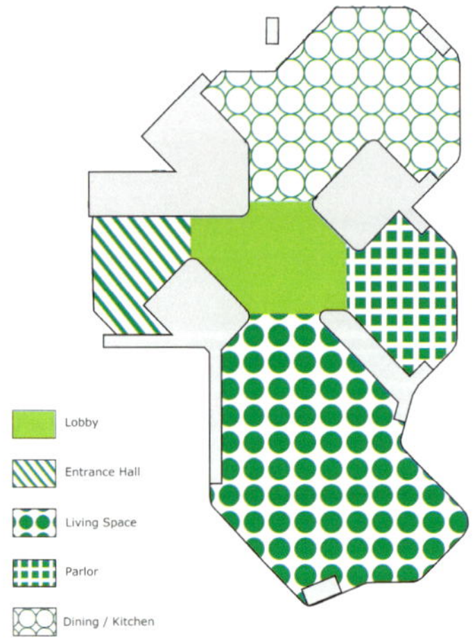

Diagram/示意图

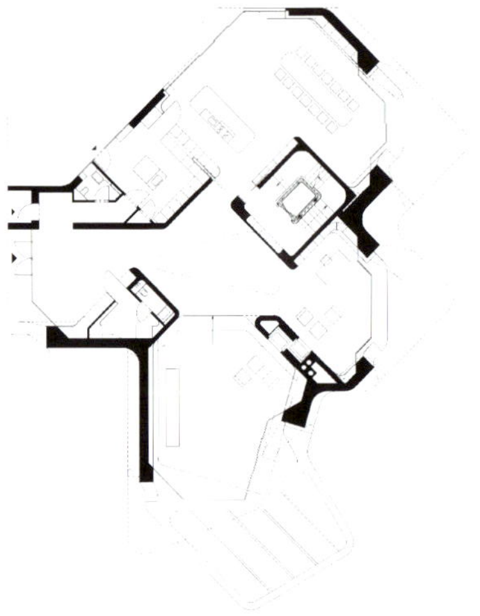

First Floor Plan/一层平面图

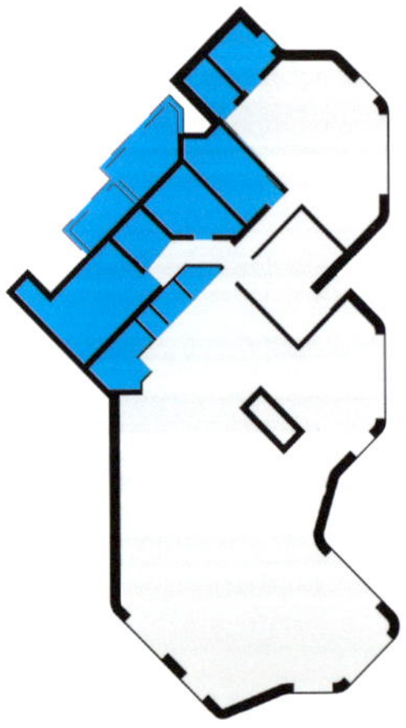

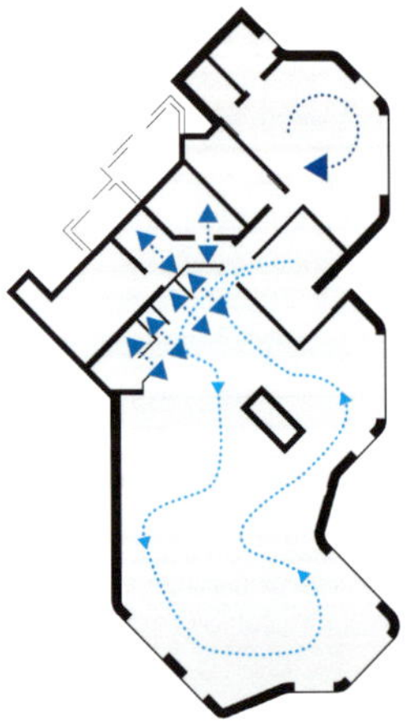

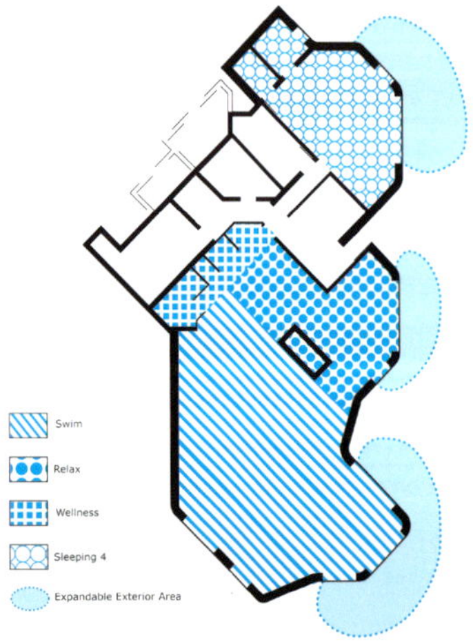

Diagram/示意图

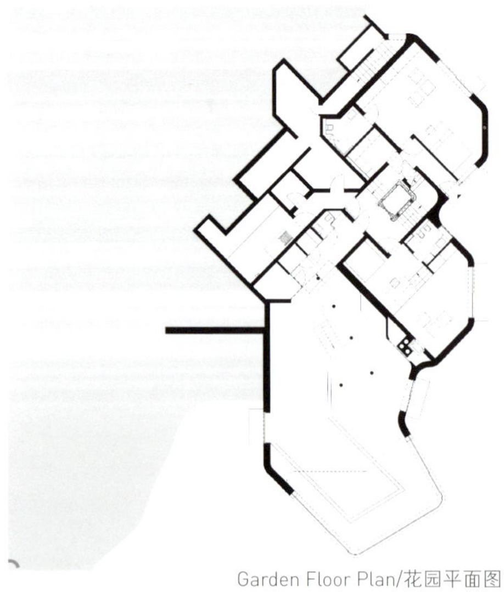

Garden Floor Plan/花园平面图

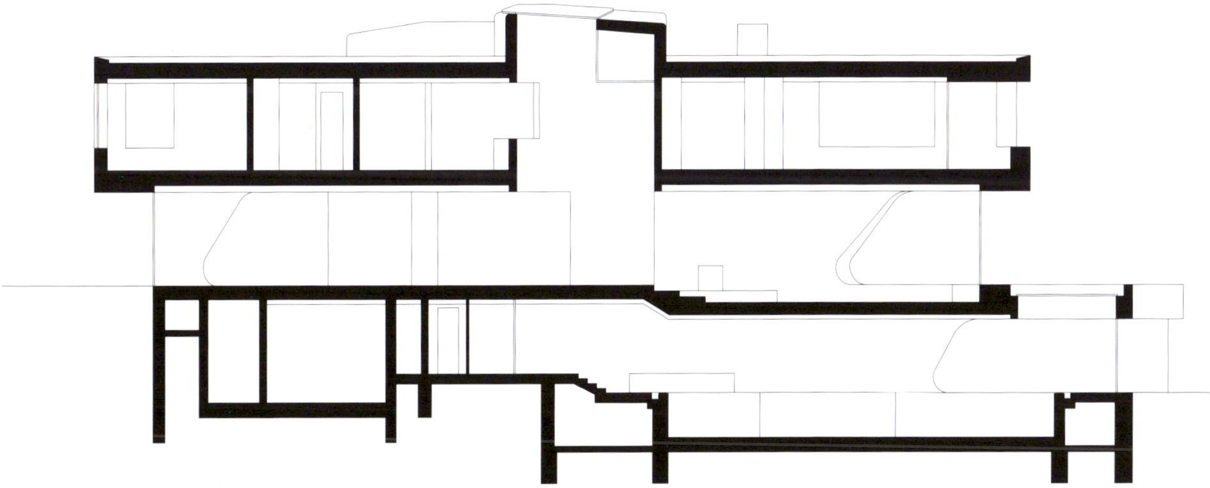

Section/剖面图

Park Avenue South

studio octopi

■ Architecture Design /建筑设计: studio octopi ■ Project Architect/项目建筑师: Chris Romer-Lee, James Lowe ■ Location/地点: London, United Kingdom ■ Area/面积: extension 20 m²
■ Completion/建成时间: 2009 ■ Photograph/摄影: Lyndon Douglas Photography

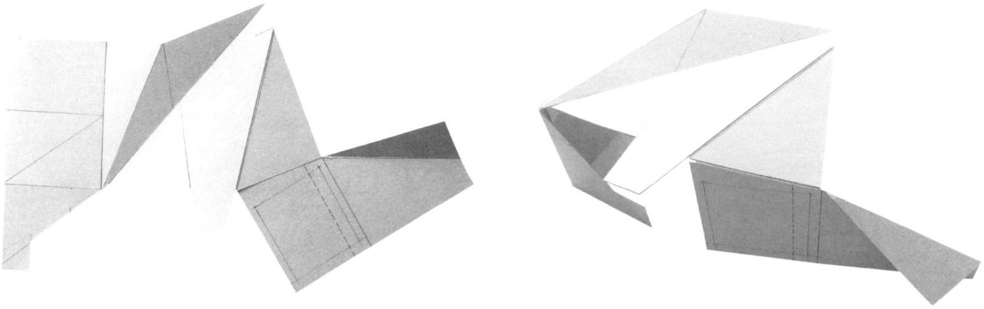

The extension to this Victorian end of terrace house is located between Crouch End and Muswell Hill. Three years in the making, the client committed to building the project one week before the collapse of Lehman Brothers. One week later, and the project may never have been realised.

The original builder was also the house's first resident, and made the most of his triangular plot by allowing the side of the building to fan out to meet the line of the adjacent public footpath. On the ground floor this resulted in an additional fillet of space splitting the living and dining rooms. It was the divisional nature of this space (used as a utility room) that the client asked studio octopi to resolve. By relocating the utility room, the plan was reordered and paved the way for an extension that linked the living spaces.

The design was developed through a series of folded paper sketch models exploring the nature of the triangular plot, the geometry and aspect. The lines of the roof ridges were drawn out from two points on the rear wall of the house, whilst the elevations extend the lines of the living room and the external rear wall of the kitchen. The structure is clad entirely in blackzinc, with standing seams tracing a path across the roof, emphasizing its complex topography and echoing the folds created in the paper concept models.

From a distance, the structure reads as a strong geometric form that has grown out from the back of the house, but at closer quarters, its edges appear to soften and the malleability of the zinc and the very slight billows in its surface come into focus. The impression formed is of a tailored garment turned inside out to reveal a complex structure of pleated seams.

Internally, the smooth planes and sharp facets of the ceiling recall an origami paper lantern, neatly folded and then popped up into three dimensions to form a bright lining to the dark fabric over-garment. Seemingly in constant motion, the planes shift and tilt, alternating with triangular roof lights that frame views of the sky, trees and distant chimney-tops.

这栋维多利亚式的联排别墅一端的加建项目，位于蹲尾区和麦斯威山之间。三年来的时间，客户致力于建设该项目，而一个星期后，雷曼兄弟倒闭。由此，该项目可能永远也不会实现。

该建筑原始的建造者也是这栋房屋的第一个住户，他充分利用起了这个三角形地块，使房屋的外围成扇形展开，紧贴公共区域。这样一来，一楼的起居室和餐厅之间就出现了多余的空间而使它们彼此分开。在原来储物室的位置设置隔断，解决了客户急需Octopi设计工作室处理的一大难题。移走了储物室，设计方案这才得以重新规划，这为日后起居空间的加建铺平了道路。

为了探寻三角地形的特性，设计师们做出了一系列的折纸模型，在房屋后墙上的两个部位，设置了屋檐。该结构被整体涂上了黑色的锌层,而接缝处则形成了横跨屋顶的一道道褶皱，这样的设计强调了该地复杂的地形，并且呼应了折纸模型上的折印。

从远处望去，该结构像是从房子后面长出的一个巨大的几何体，但是近处看来，它的边缘似乎弱化了锌层的僵硬感，形成一种褶皱感。

该建筑仿佛是将一件量身定做的衣服的里面翻了过来，露出了精细缝合的内部构造。室内方面，平滑的屋顶以及锋利的切面，使人不由得联想到折纸的式样。简洁的折叠方式，却能立即呈现出三个不同的空间，这与外面黑色的镀锌层形成鲜明的对比。这些平面伴随着三角形屋顶变换的灯光，不停地移动、旋转，将天空、树木以及高处的烟囱一幕幕地定格下来。

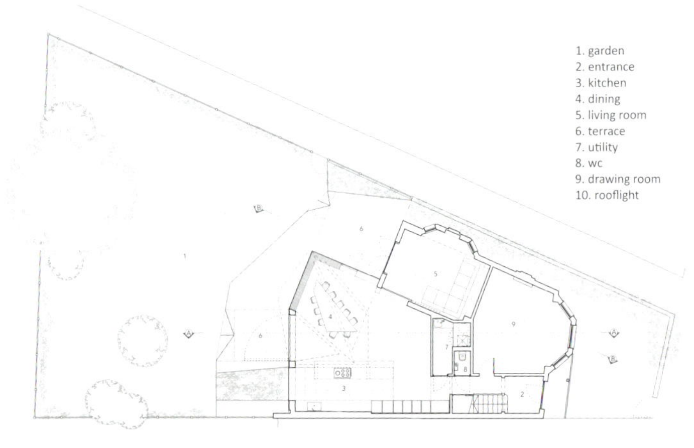

1. garden
2. entrance
3. kitchen
4. dining
5. living room
6. terrace
7. utility
8. wc
9. drawing room
10. rooflight

Ground Floor Plan/底层平面图

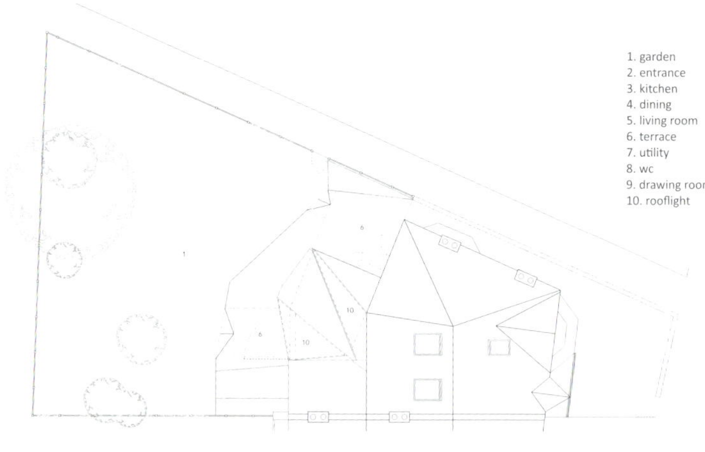

1. garden
2. entrance
3. kitchen
4. dining
5. living room
6. terrace
7. utility
8. wc
9. drawing room
10. rooflight

Roof Plan/屋顶平面图

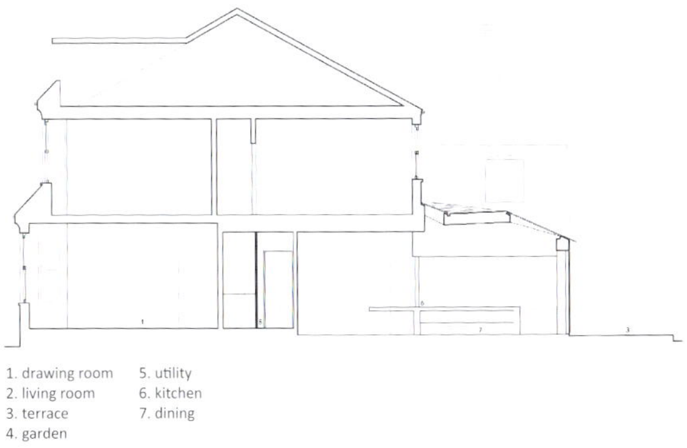

Section/剖面图

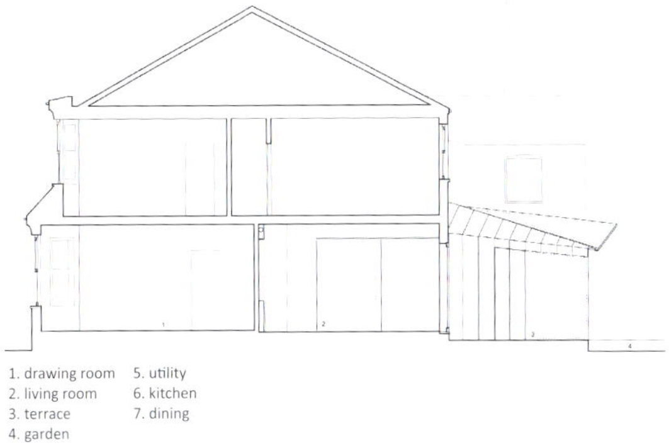

Section/剖面图

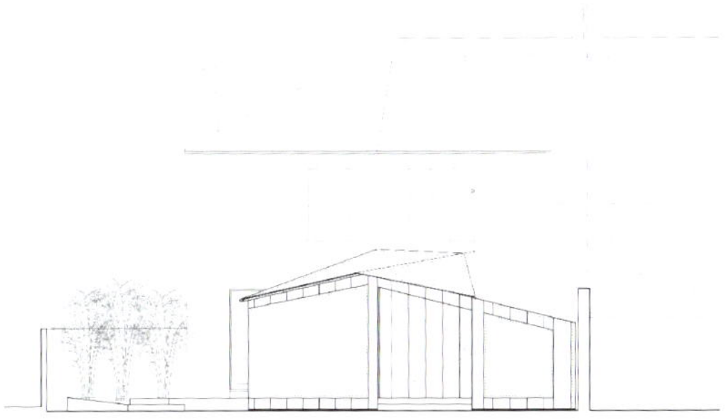

Facade/立面图

Casa Son Vida

TecArchitecture

■ Architecture Design/建筑设计: TecArchitecture ■ Project Architect/项目建筑师: Sebastian Knorr ■ Interior Design/室内设计: Marcel Wanders Studio ■ Developer/开发商: Cosmopolitan Estates
■ Location/地点: Mallorca, Spain ■ Area/面积: 800 m^2 ■ Completion/建成时间: 2009 ■ Photograph/摄影: Gaelle Le Boulicault, Marcel Wanders Studio

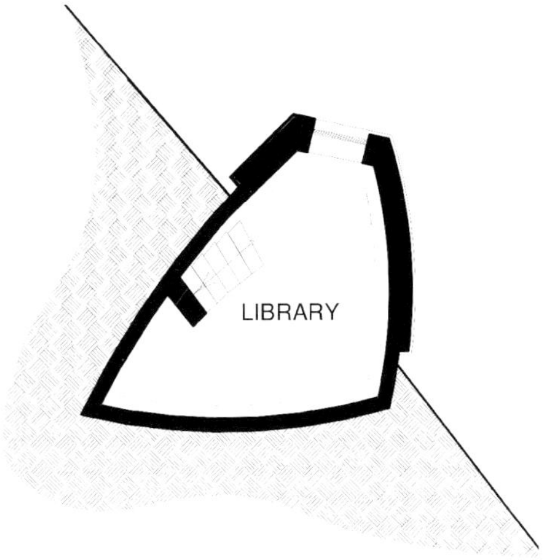

Located approximately 15 minutes outside of Palma de Mallorca in an exclusive community called Son Vida, the eponymously named Casa is an example of design poetry in motion. The completion of this remarkable home marks the arrival of third millennia architecture on the island of Mallorca. The task at hand was straightforward: take a very un-extraordinary existing Mediterranean villa constructed in the 60s and transform it into something extraordinary. The 800m^2 luxury villa transgresses the constraints of site and context, redefining luxury architecture as it is typified by the traditional and prolific Mediterranean and Tuscan styles otherwise found on the island.

As the first villa to be completed as part of the Platinum Estates, Casa Son Vida is a hint of what is to come. This property is not for sale.

该项目在距离Palma de Mallorca市15分钟车程的一个独立社区中，该社区名为Son Vida，而与这个社区同名的Casa Son Vida项目，是将设计的艺术性融入动态模式的一个很好例证。这个杰出建筑的竣工标志着旗下著名品牌第三千年正式入驻马洛卡岛。眼前的这个任务很明确：把一个建于20世纪60年代平淡无奇的地中海风格的别墅改造成为一个别具特色的建筑。800平方米的奢华别墅早已超越了原有地形和环境的限制，设计重新定义了豪华的建筑，因为它结合了传统而丰富的地中海建筑特色，体现了典型的托斯卡纳风格。

作为Platinum Estates公司开发的第一幢别墅，Casa Son Vida项目是对将来建筑风格的探索。因此，这一别墅不会作为商品房而出售。

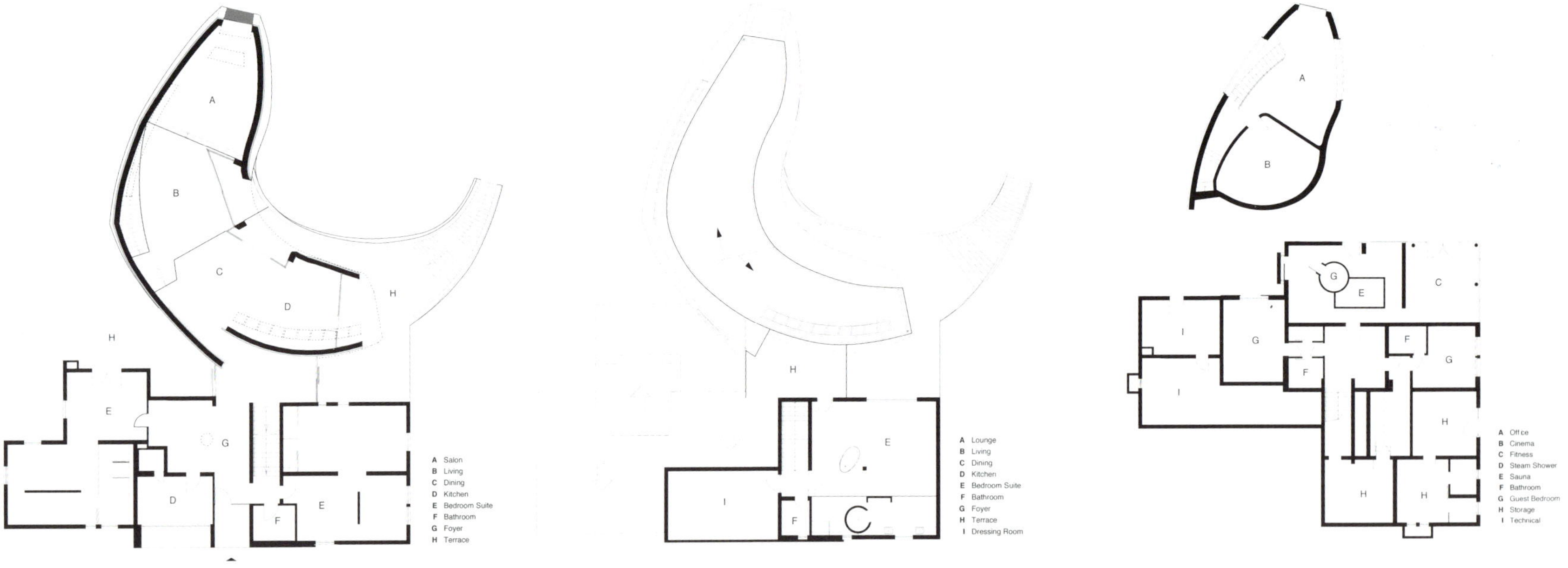

A Salon
B Living
C Dining
D Kitchen
E Bedroom Suite
F Bathroom
G Foyer
H Terrace
A Lounge
B Living
C Dining
D Kitchen
E Bedroom Suite
F Bathroom
G Foyer
H Terrace
I Dressing Room
A Office
B Cinema
C Fitness
D Steam Shower
E Sauna
F Bathroom
G Guest Bedroom
H Storage
I Technical

Plan/平面图

TecArchitecture HQ

TecArchitecture

- Architecture Design/建筑设计: TecArchitecture
- Project Architect/项目建筑师: Sebastian Knorr
- Location/地点: Ermatingen, Switzerland
- Area/面积: 900 m^2
- Completion/建成时间: 2008
- Photograph/摄影: Peter Allgaier

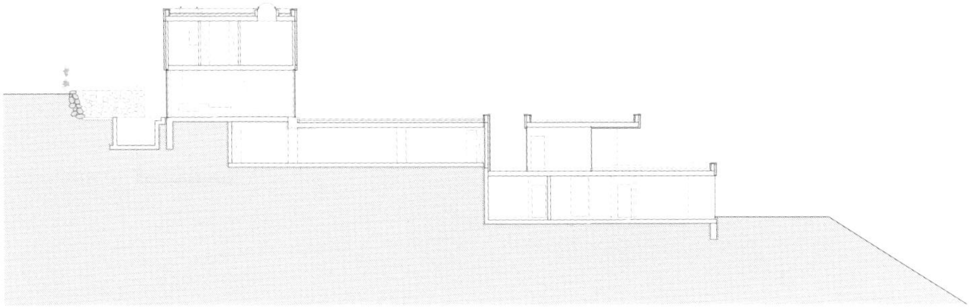

This project is a compound of spaces on lake constance, recently finished by tecARCHITECTURE swiss AG/ tecDESIGN los angeles. Tec's new offices are part of the building as well as the residence of tec coo heiko ostmann and his family.

We decided to come up with a steel/ concrete construction to maximize thermal storage, and guarantee large spans, in order to pull the outside and the breathtaking views into the building. All units share that view. Our office on the lowest level is a simple exposed concrete loft. The building faces north, so the solar impact on the glass is negligible. The south elevation is almost completely closed-it also faces the local cemetry...

The structure Is roughly 900m² and also contains some underground parking accessible by car lift. Heiko, who lives in the house with his family, and me have been with the first group of architects to address issues of sustainability in -then- germany. We designed some of the first naturally ventilated low - impact office buildings, and have carried those ideas through our work.

The ermatingen building features state of the art sustainable technologies. Energy is being generated geothermally (we drilled 800 ft down), and solar. Heating and cooling works through activated masses - hence the concrete, open floors etc.

该项目由瑞士设计公司TecArchitecture设计完工，是在康士坦茨湖上的一个复式建筑。这个建筑的一部分是Tec新的办事处，一部分是Tec首席执行官一家居住的地方。设计师构想设计一种钢结构或是混凝土结构的建筑，这个建筑能够将储热器设计为最大规格，能够保证较大的跨度，还能将室外的美景尽收室内，使每间卧室都能享受到这份美景。办公部分在建筑的最底层，是一个显眼、简易的混凝土阁楼。阁楼面朝北面，所以太阳光对阁楼的影响几乎可以忽略不计。南立面几乎是完全封闭的，因为它正对着本市的公墓。

该建筑面积大约为900平方米，另外还附有地下停车通道。住在这儿的海科和该建筑师本人，是第一批解决了当时德国可持续发展问题的建筑师。他们设计的建筑，最早实现了自然通风换气，降低了建筑对环境影响，并且他们把这些理念运用到日后的设计中。

该建筑的特色在于艺术地运用了可持续发展技术。为了能够利用地热能源供暖，建筑师向地下钻井约244米。地热通过地下的启动装置，传导至混凝土以及开放的地板，保证了室内的供热与制冷正常运转。

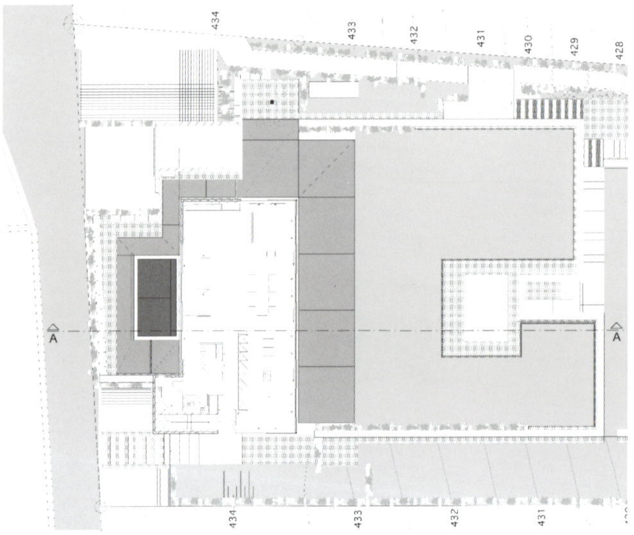

Third Floor Plan/三层平面图

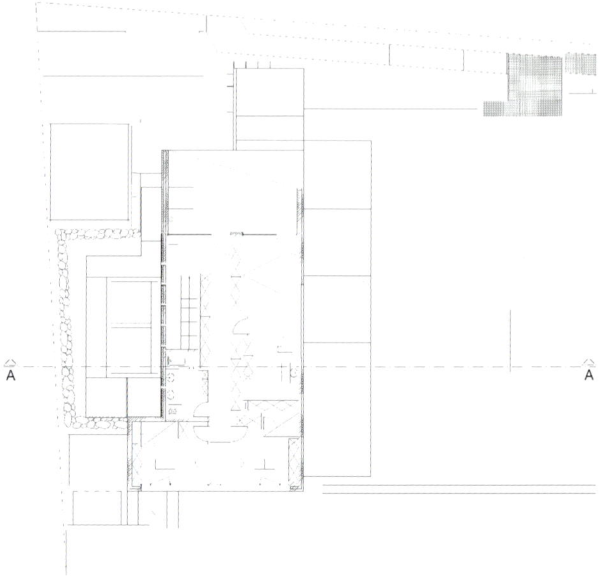

Fourth Floor Plan/四层平面图

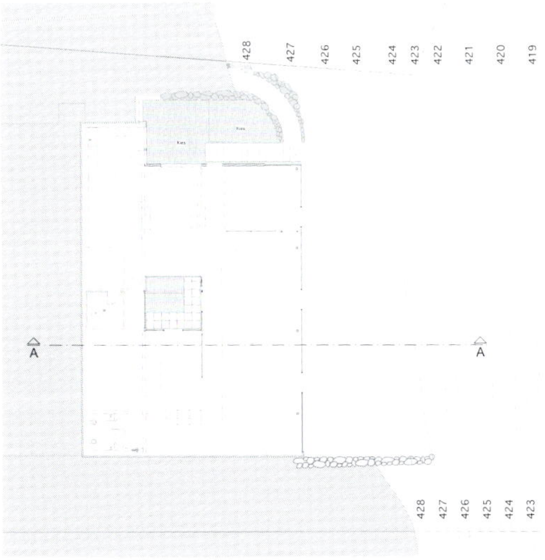

First Floor Plan/一层平面图

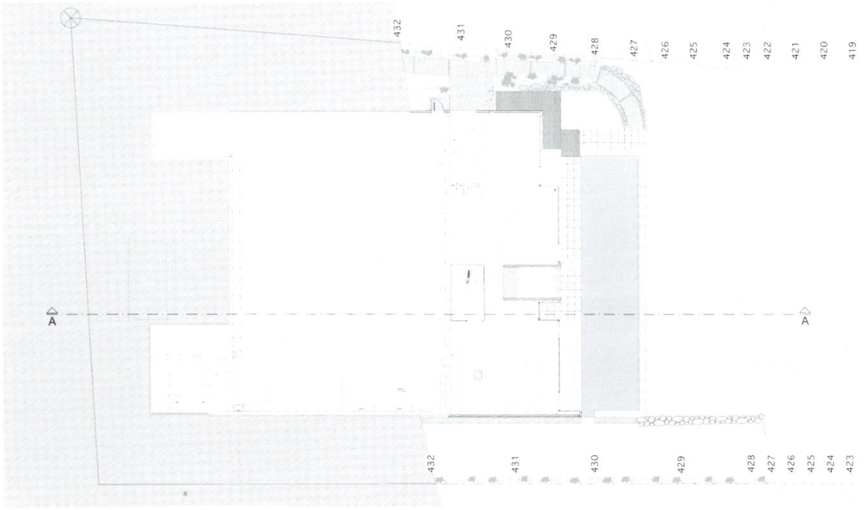

Second Floor Plan/二层平面图

House M

Caramel architekten

■ Architecture Design/建筑设计: Caramel architekten ■ Project Architect/项目建筑师: Caramel architekten ■ Location/地点: Austria ■ Area/面积: 108 m² ■ Photograph/摄影: Otto Hainzl

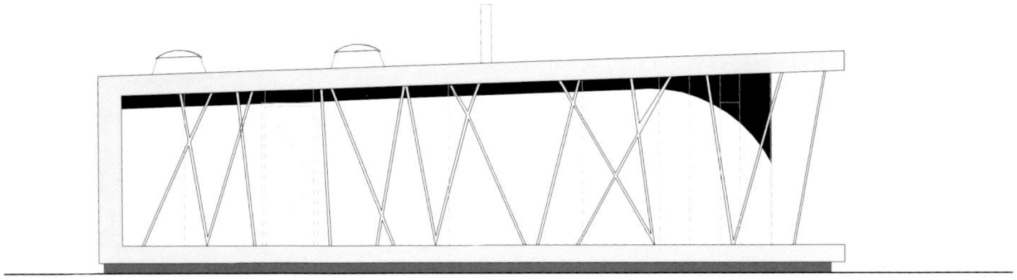

Sitting on the slope of pöstlingberg in linz, the 12 x 12 meter cube takes full advantage of the lot by establishing itself as far into the north-eastern corner of the site as building codes permit. These two sides have a very hermetic appearance, thus preventing neighbors from looking in.

By contrast, the house opens itself toward the south and west with two all-glass façades. The west side of the building offers a magnificent view of the danube valley and receives sunlight until late in the evening; an externally mounted textile curtain guards against excessive sunlight and overheating. The south façade pulls away from the edge of the cube In free-form retreat, producing a covered terrace area. Extending directly from the terrace is a large swimming pond.

Moving inwards from the southwest open kitchen / dining / living areas that face the swimming pond, one comes to the non-public bedroom units lying to the northeast.

The cellarless house is made out of prefabricated high-performance structural insulated panels, which were installed onto the concrete floor slab in only a few hours with the help of a truck-mounted crane. The diagonally arranged pillars along the west façade provide structural reinforcement.

Slightly inclined toward the east, the roof slab offers gutterless drainage over the façade and into a gravel-filled french drain at the back of the house.

The cube is enveloped in synthetic membranes. For the roof area white epdm was used; for the façade area (invisibly welded to the roof liner), white teflon-coated pvc, which was secured at intervals through the thermal insulation layer between the outer membrane and the structural panels.

The simple form of the building volume, the high degree of prefabrication, its excellent cellulose-fiber thermal insulation, as well as the application of calorific value technology and external shading all contribute to making this a very economical and ecological building.

该建筑位于林兹市中伯斯特林山的山坡上，一个边长为12米的正方形建筑完全占据了这块地皮。依照建筑规范要求，该建筑物在设计上最大限度地向东北方向延伸。设计师将东面和北面设计为封闭式的墙壁，以保证该建筑的隐秘性。

与此相反，这栋房子的西面和南面均为开放式的落地玻璃墙。从房子的西面可以看到多瑙河壮丽的景色，一直到很晚，这里都能被太阳照射。一对通顶的丝质窗帘，遮挡了外面刺眼的阳光和热气。而在房子的南面，沿着建筑结构的边缘向南扩展了一块结构随意的僻静处所，在此增加了一个遮阳的平台，面向平台处还开辟了一个巨大的室外泳池。

房子的西南侧是开放式的厨房、餐厅和客厅，面朝泳池。从这里向里走，人们可以看到东北方向上的一个个私密的卧室套间。

这栋半包围式的房子是由高质量的预制保温板建成的，所选的材料，在吊车的配合下只需要个把小时就可安装好。设计师为了强化和巩固房屋结构，在房子西侧的对角处还设计了一个立柱。

房子的屋檐稍稍向东倾斜，因为在屋顶处设置了排水系统。排水管跨过墙面转到房子的背面，最终进入石子路下的法式排水管。

立方体的建筑外部贴有一层人造薄膜。屋顶外部使用的是三元乙丙橡胶，封闭式的墙面使用的是聚四氟乙烯材料。这些材料可以通过隔热装置间隔地将人造外膜和内部结构板保护起来。

这栋房子不仅构造简单，预制材料质量上乘，而且配备了良好的光纤隔热装置并采用了先进的发热技术和巧妙的户外遮阳设计。所有这一切将这里打造为一个既实用又环保的理想住宅。

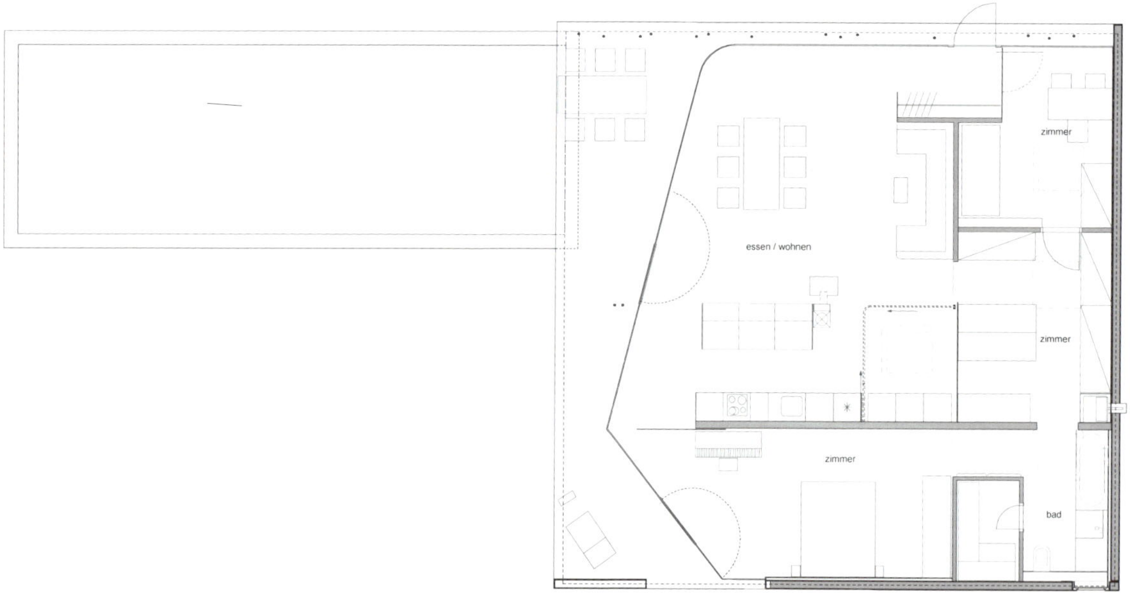

Plan/平面图

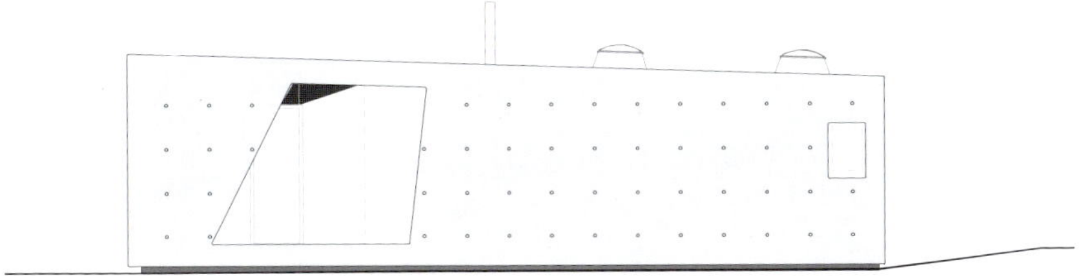

Facade:East/东立面图

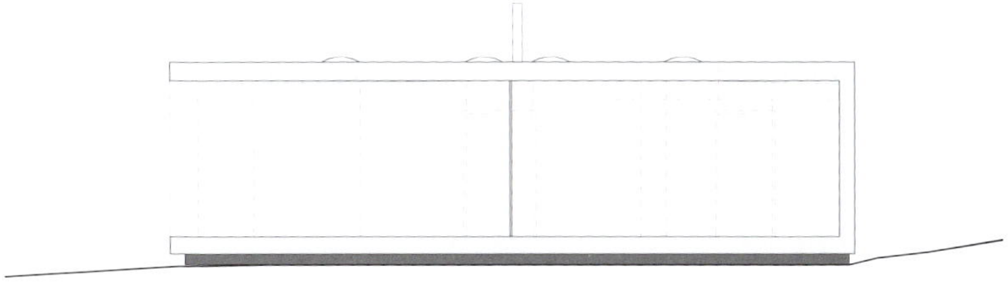

Facade:South/南立面图

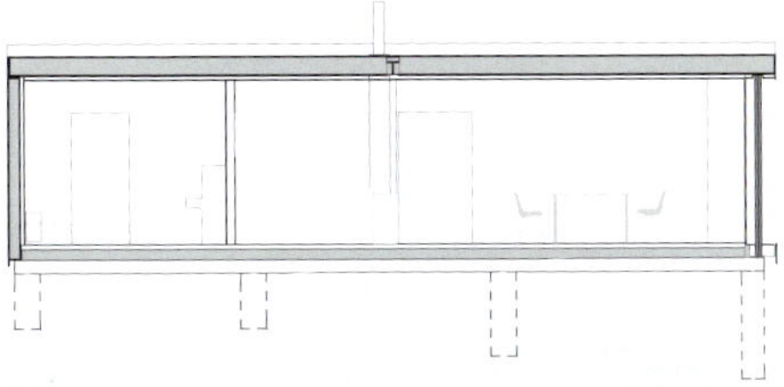

Section/剖面图

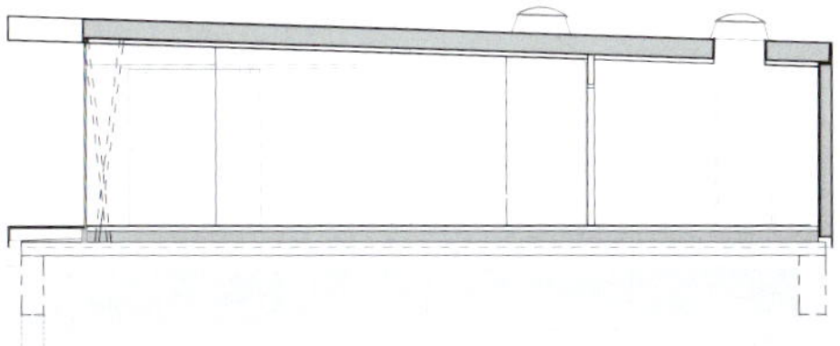

Section/剖面图

House P

Caramel architekten

- Architecture Design/建筑设计: Caramel architekten
- Project Architect/项目建筑师: DI Barbara Sabine Bovelino (project leadership and interior design)
- Location/地点: Austria
- Area/面积: 130 m²
- Photograph/摄影: Hertha Hurnaus

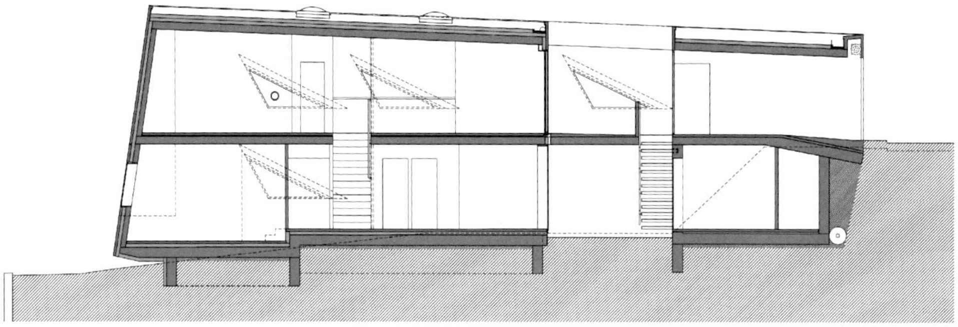

Like a curious animal that might have been formed from Tetris blocks, House P seems to gaze down at the valley from its perch on this sloping property in Klosterneuburg. The seamless impression of the homogeneous sculptural form is strengthened by the use of prefabricated concrete elements in the façade. The access way on the street side follows the structure's longitudinal axis, passing the garage and terrace and entering the house, where it pushes past the dining room/kitchen area, cloak room, and stairway landing, and finally disappears out the glazed front of the living room and into a breathtaking view of the Danubian plains. The length of the building is interrupted only by the transverse orientation of the pool, which extends as a visual continuation of the lower covered terrace. Here at the lower garden access level there are two bedrooms, two bathrooms, a sauna, and a small studio that opens out on the garden and the lower covered terrace. Located across from the studio on the other side of the terrace is the cellar, which is inserted into the slope beneath the garage.

Depending on its orientation, the façade is open or hermetic. While the west side with the dining room/kitchen and terrace opens out fully onto the pool and garden, the east side, which faces the nearby neighbors is almost completely closed off. Here, gill-like flaps direct one's view toward the valley and create a private sphere for the living areas. By contrast, half the façade on the north side of the living area is thrown open, affording a complete view of the valley. This wide-open impression is Intensified by the flush-glazed façade that extends overhead, becoming part of the roof.

该住宅的主人看中了克洛斯特新堡的山谷，决定把房子建在山谷顶端的一片空地上。House P看似是一座由俄罗斯方块堆砌而成的奇特物质，该建筑墙面的预制混凝土材料加固了这个雕塑般的结构，使建筑设计达到了完美的效果。进入院子的通道临街而设，沿着房屋结构的纵向轴心行进，可穿过车库和平台，来到室内，再经过餐厅和厨房、更衣室、楼梯台阶和光鲜的客厅后，映入眼帘的是多瑙河平原的壮丽景色。建筑物被横向的游泳池拦腰截断，而游泳池仿佛是低处平台的延伸。在这块较低的通向花园通道的两侧是两间卧室、两间浴室、一间桑拿房和一间小型工作室，工作室面朝花园和较低的遮顶平台。

根据不同的朝向，房屋的墙面分别被设计为开放式和封闭式两种。朝西的餐厅、厨房和平台完全敞开，面朝着游泳池和花园；而朝东的房间由于紧邻邻居住所，因此基本上是全部封闭的。在这里，窗口的设计很特别，像一片片的鱼鳃，不仅能够看到远处的山谷，也避免了客厅隐私外露的风险。与此相反，客厅朝北的半面墙壁是向外开放的，这样主人能够尽情饱览山谷的美景。设计师利用光鲜的墙面，在视觉上增强了建筑物的宽度，形成开放效果，使这种敞亮感超过墙面，越过头顶，充满半个屋顶。

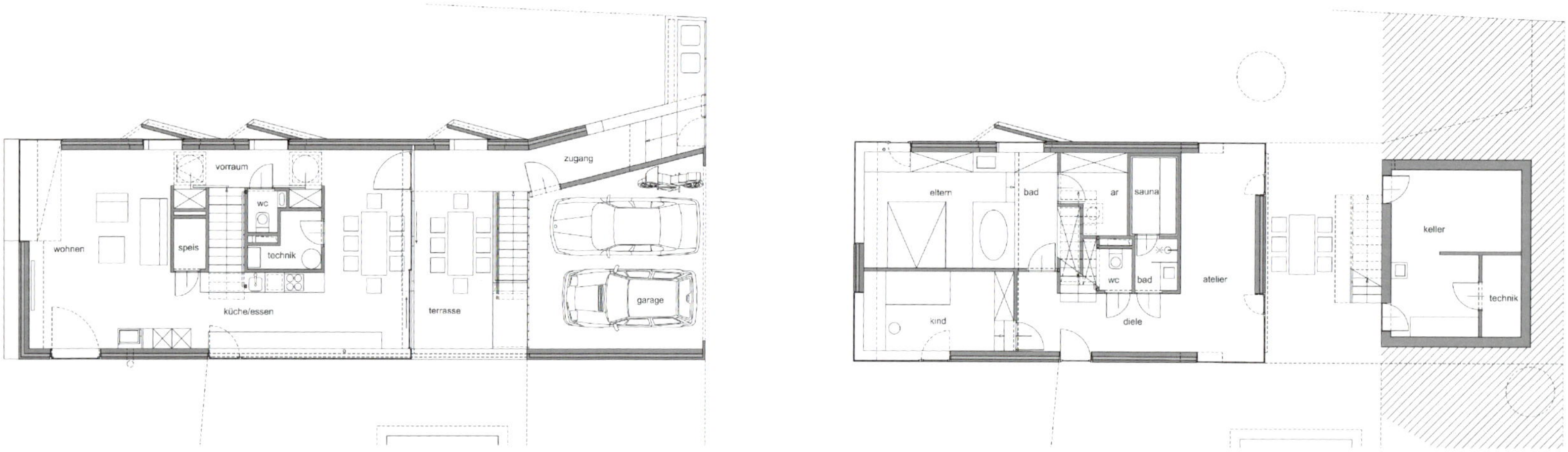
vorraum
zugang
wc
wohnen
speis
technik
küche/essen
terrasse
garage
eltern
bad
ar
sauna
wc
bad
atelier
kind
diele
keller
technik

Plan/平面图

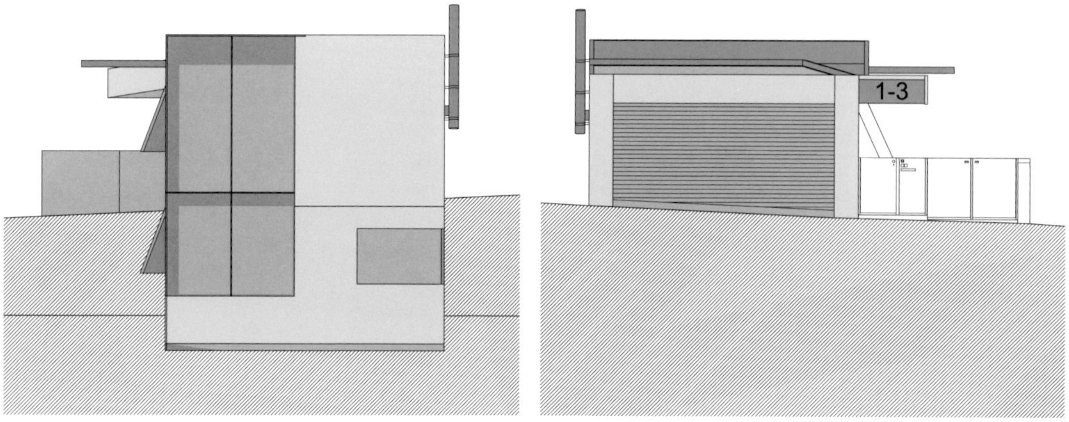

Facade: North/北立面图

Facade: South/南立面图

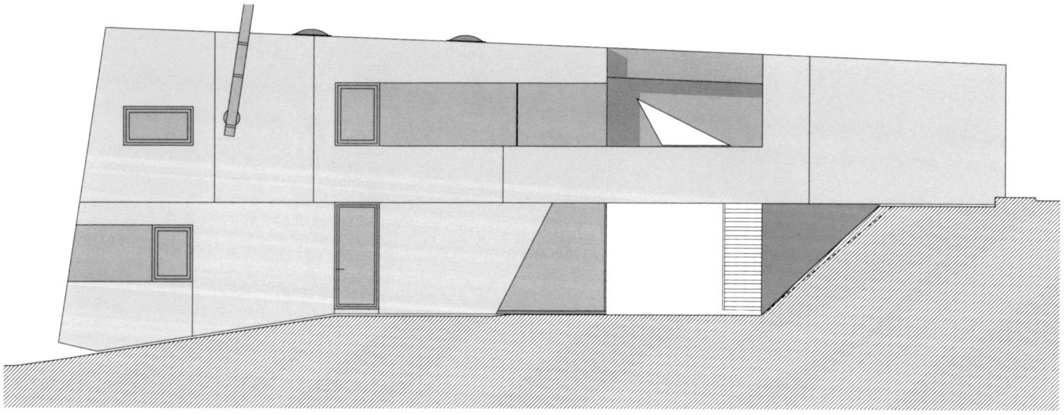

Facade: West/西立面图

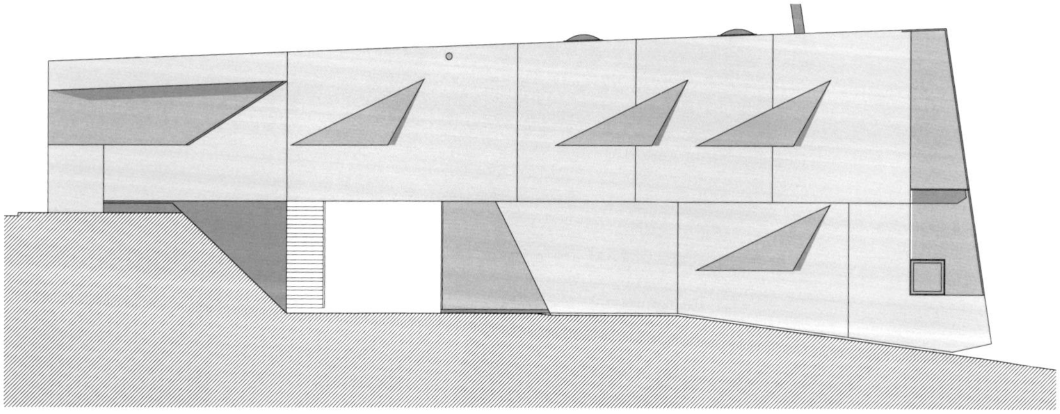

Facade: East/东立面图

Jas & Elisa

Atelier Vens Vanbelle

■ Architecture Design/建筑设计: Atelier Vens Vanbelle
■ Project Architect/项目建筑师: Dries Vens, Maarten Vanbelle
■ Location/地点: Veldegem, Belgium
■ Area/面积: site 528 m²
■ Completion/建成时间: 2009
■ Photograph/摄影: Atelier Vens Vanbelle

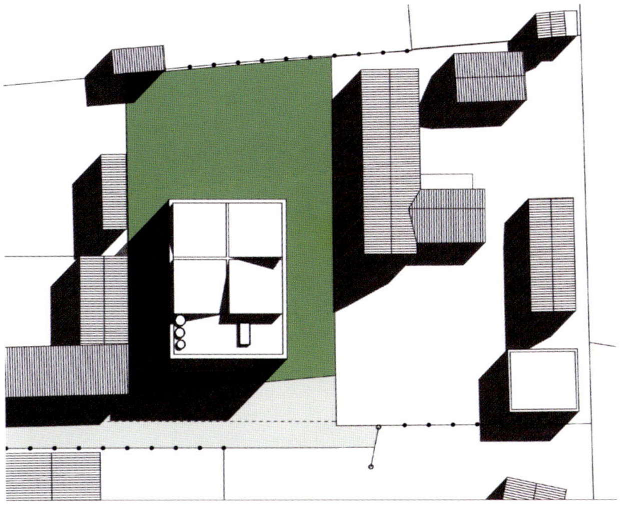

The design of this house was determined by two striking factors: first, the plot is enclosed by other plots, and secondly the clients plan to have lots of children.
The plot can be accessed via a narrow road which leads to an enclosed piece of land adjacent to the backs of the gardens of the neighbors. As it is a Flemish tradition to have a pavilion at the end of your parcel, the building plot is surrounded by backsides of these little pavillions.
In the design we attempted to provide an answer by looking at the facade as if it were also a kind of 'backside' towards the neighboring parcels. This was translated into a volume that opens itself towards the parcel only via a few strategically chosen openings on the ground level. The first floor is kept almost completely closed. We created a number of small openings from which you can peep inside the gardens of the neighbors, but the neighbors can't look into the house. Daylight on the first floor is provided by the roof, which has a specific form and is sepparated into four parts.

To strenghten the idea of the 'backside' and to break the massiveness of the masonry we have chosen to incorporate a pattern of japanese wallpaper into the facade.
The clients also have a great love for children and cherish the desire to once have a family with four kids. Although we thought it would be absurd to build a house with four rooms for children at a moment there were no children, so we looked for a decent alternative.
This was translated into some 'sleeping boxes', of which the principle is very simple: on the first floor we created a large open space that can be filled according to the needs of the moment. Each time a child gets born the clients can add a 'sleeping box'. In this way they can not only spread the construction budget over the years, but it is basically incidental how many children eventually will come.

这栋房子的设计需要考虑两点比较突出的因素，第一，这块地皮位于其他地皮中间；第二，委托人希望生养很多个孩子。

可以通过一条小路进入这块地，这条小路一直通向隔壁邻居家的花园后墙。由于弗兰德人的传统是要在自家的院子后面建造一个凉亭，这样就不难理解为何该址周围布满了这样一个个的小凉亭。
在进行房屋设计时，设计师考虑将建筑表面设计风格与邻居家后院的建筑风格相统一。通过在底层巧妙地设置一些窗口，使该建筑只面向自家院落。房子的首层基本上是封闭的，设计师只安放了一些很小的窗口，这样既可以欣赏到邻居家的花园，又为自己的房子增加了隐秘性。屋顶有着独特的造型，它被分成了四个部分，日光透过屋顶可以照射到首层的房间中。
为了突出“后院”的概念，同时打破砖石的厚重感，设计师选用一款日式的墙壁，使其成为建筑物立面的一部分。
客户非常喜爱孩子，并且特别希望能够组建一个六口之家。尽管为一个目前还没有孩子的家庭设计四间儿童卧室显得有些荒唐，但是设计师找到了一个合理的替代方式。
这个办法就是构想设计出一些 “睡觉的盒子”，其建筑思路非常简单：在首层设计出一大片开放式空间，客户可根据需要来设置。这样，每当有一个宝宝降生，客户只要增加一个“睡觉的盒子”就可以了。这种设计方式不仅节省了建造成本，而且也满足了即将到来的宝宝们的需求。

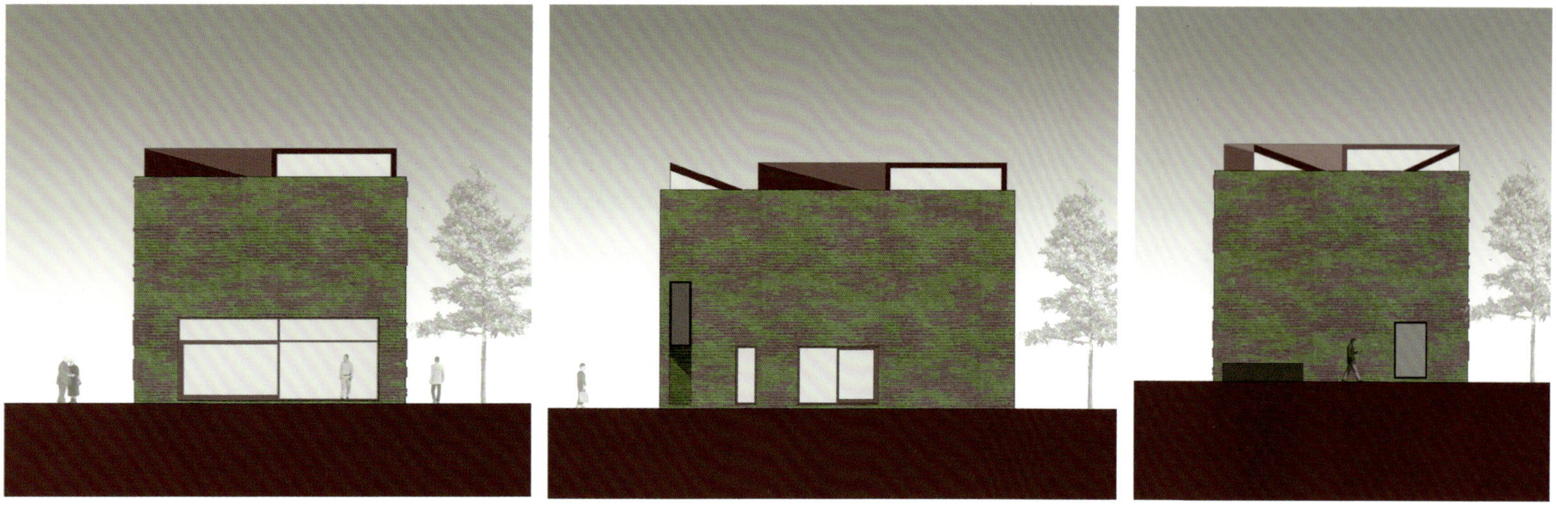

Facade/立面图

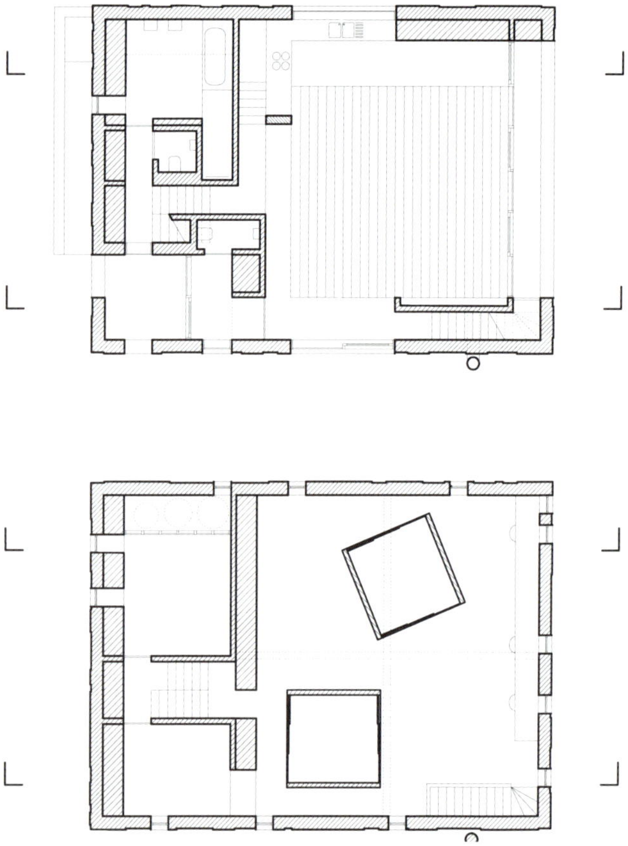

Plan/平面图

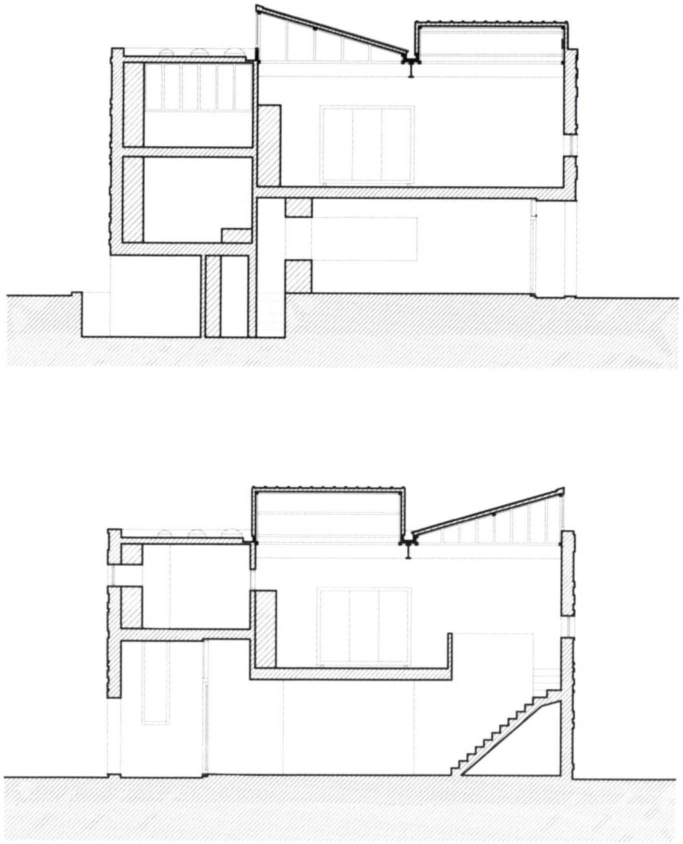

Section/剖面图

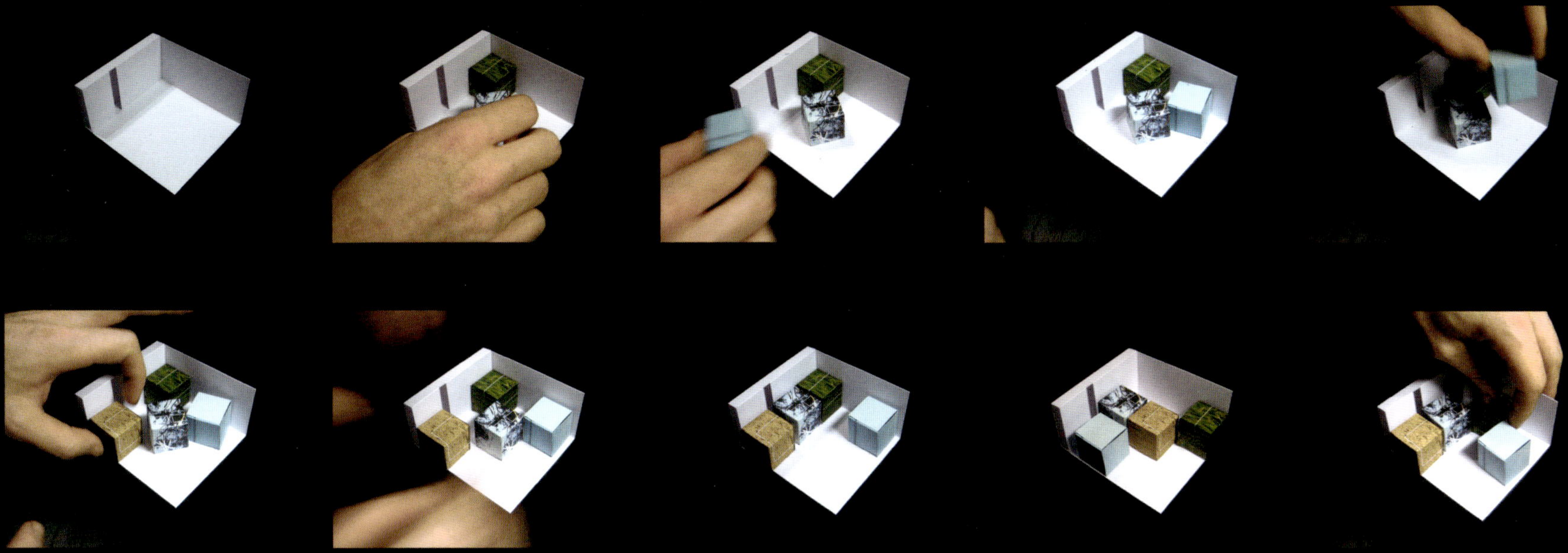

Q House

asensio_mah

■Architecture Design/建筑设计: asensio_mah ■Design Team/设计团队: Leyre Asensio-Villoria, David Mah, Diego Repiso, Jennifer Chuong, Kaizen Chen, Jon Aguirre ■Location/地点: Spain
■Area/面积: site 900 m^2/building 142.59 m^2/gross 320 m^2 ■Completion/建成时间: 2009 ■Photograph/摄影: Ricardo Loureiro

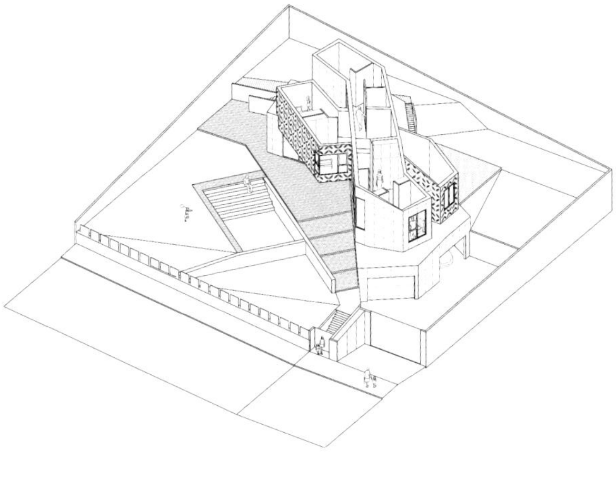

The house is a conscious exercise in developing an alternative domestic environment to the surrounding villas of the new suburban neighborhood. The solutions for the development so far have typically been compact villas located on abruptly leveled gardens, irrespective of the complex topographical condition of their sites. Our ambition for producing an alternative domestic atmosphere is developed by constructing a more explicit relationship between the house and garden with the existing conditions of the steep site. This organizational strategy for the house sought to register the difference in topography within the parcel by organizing a series of terraces that configure the framework for a landscape with differentiated characters. This deliberate geometric configuration affords multiple readings of the outline of the house while facilitating a rich experiential lifestyle within its volume and landscape. Specific organizational and material strategies were developed to produce different volumetric and perceptual readings that change with the different vantage points towards and within the house.

The building is organized in three bands that are arranged around a central circulation core. These three bands maintain a prevailing orientation in the northeast-southwest direction to secure maximum daylight in every room. While the bands configure and organize the different rooms, the circulation core underpins a switchback pattern of shifting orientations with the gradual vertical movement through the house.

The house is clad in dark "composite" panels that have been customized with digital fabrication techniques. These customized panels are used to articulate sections of the house volume in order to introduce legibility to the overall form. These panels offer a range of different surface consistencies and patterns to the house that reflect the sites changing light conditions in multiple ways, producing an ever changing range of texture and tones.

在城郊新建的别墅周围营造一种非同寻常的氛围，是设计这栋房子时的一种有意识的探索。而这种探索是针对别墅修建时不考虑复杂的地形条件而将原先陡峭的花园变成平地的情况下进行的。为此，我们的目标是在现有的陡峭地势下，把建筑物与花园之间构建出一种更加清晰的关系。该房子的设计方略是保留不同的地形以及重建一系列梯地，使其成为别具特色的一处风景。设计师精心设计的几何形构造，以其宽敞的室内空间、完美的自然景观，为居住者提供了完善的生活设施的同时，极大地丰富了房屋的外观形象。通过运用独特的结构和材料，设计师使房屋内外都呈现出不同的空间和视觉感受。

该建筑物围绕着中央循环核心组建成三部分，这三部分保证了由东北到西南主要方向上每个房间的采光，确保每个房间能够最大限度地享受到日光。虽然这部分的房间格局不同，但是房子的循环核心支撑起了一个曲折的形式，使其能够在房间内根据逐渐垂直的运动变化而调整方向。

房子外表是黑色的复合型面板，是根据数字化构造技术量身定制的。这些定制的面板能够区分建筑空间的不同部分，清晰地展现建筑的整体结构，并呈现出不同的质地和样式，也反映出建筑在多种光照条件下，可以产生不断变化的气质和风格。

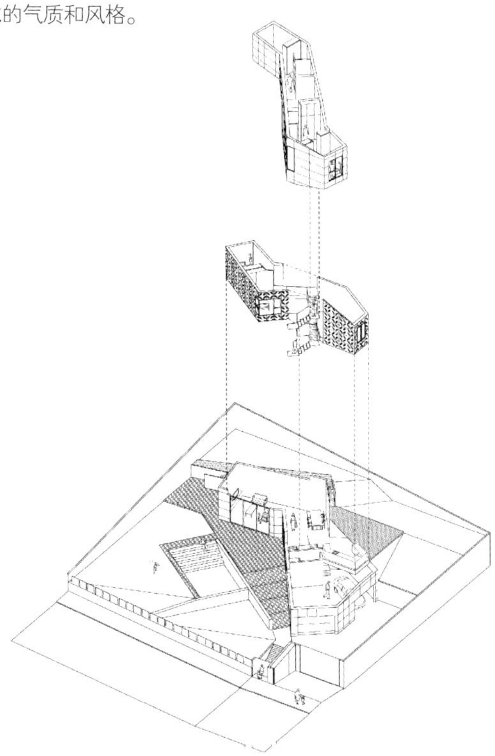

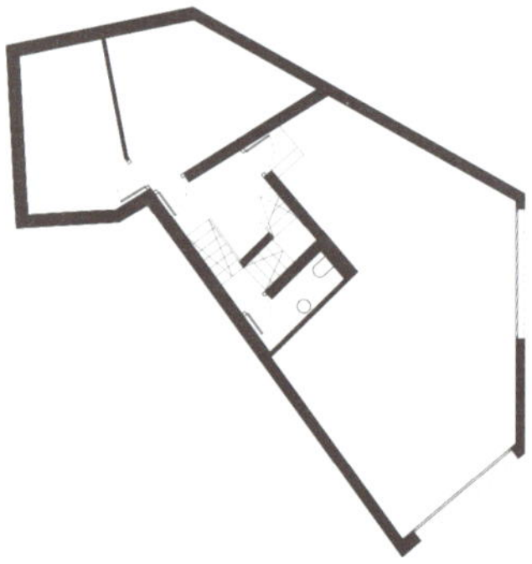

Basement Plan/地下一层平面图

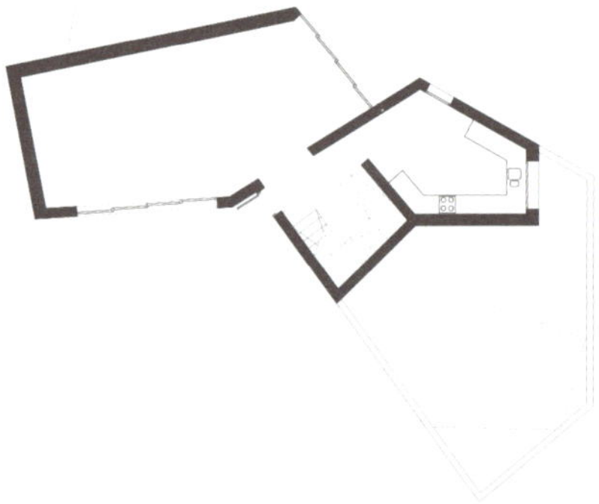

First Floor Plan/一层平面图

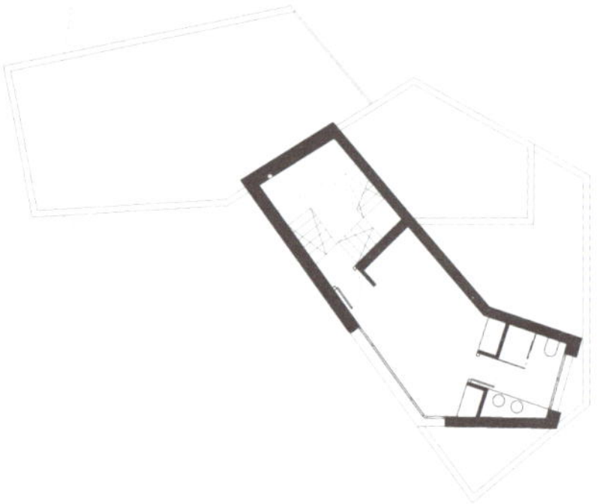

Second Floor Plan/二层平面图

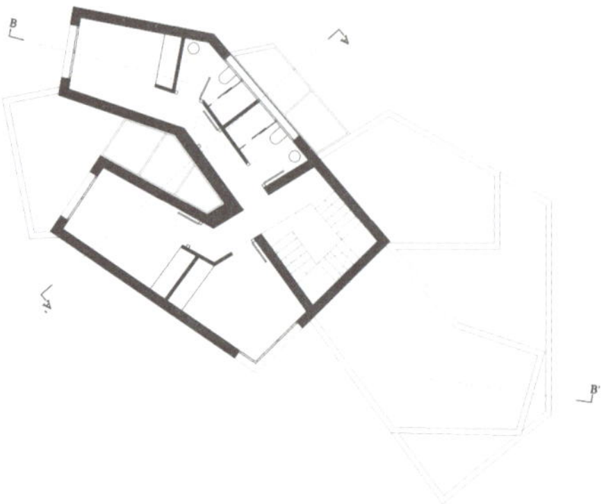

Third Floor Plan/三层平面图

Facade: East/东立面图

Facade: South/南立面图

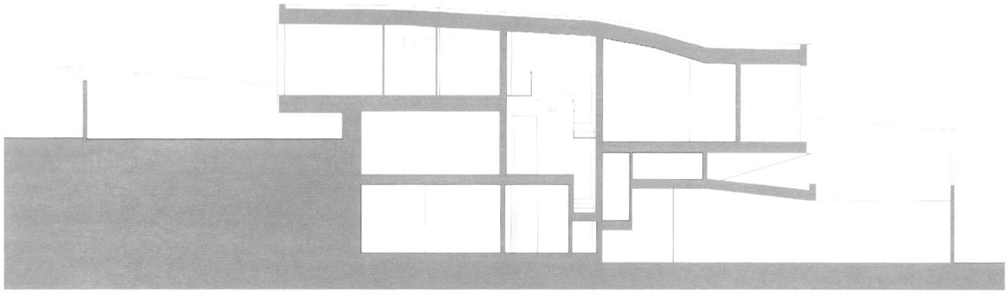

Section/剖面图

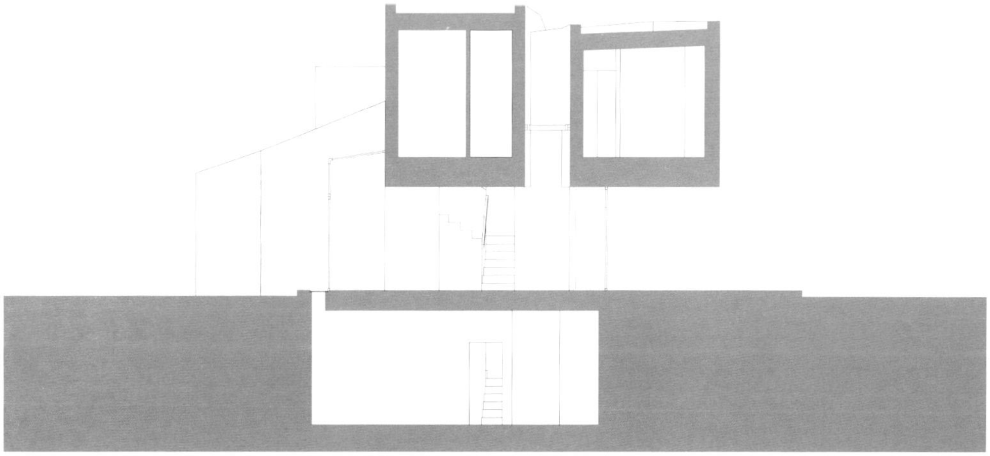

Section/剖面图

Schierle House

Matthias Benz architecture & design

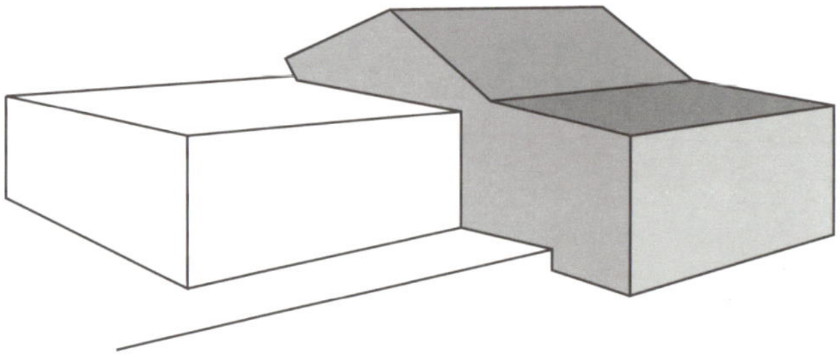

■ Architecture Design/建筑设计: Matthias Benz architecture & design ■ Project Architect/项目建筑师: Matthias Benz ■ Location/地点: Gerzen (Niederbayern), Germany
■ Area/面积: 304 m^2 ■ Completion/建成时间: 2008 ■ Photograph/摄影: Thomas Benz, Rathschek Schiefer

It was in one of these town, located at the outskirts of Munich, that a family formed by a married couple and their daughter stated the objectives for their home; that is, to build and inhabit a house, that would sensibly respect tradition, but at the same time provide an individual and contemporary style.

Their home would have to replace the remains of a previous farm, on a 2000-sq.m.-site defined by a gentle hill slope and an excellent view over the town. The strict urban regulations allowing only pitched-roof buildings challenged the client's desire for individuality. In addition, in order to avoid the rejection from the neighbors, it was important to abstain from anything that could alter the panoramic view over the valley they already enjoyed.

A simple geometry was adopted for the house, combining two solid volumes enabling not only to maintain but improve the neighboring homes view over the valley.

White is the companion for the peace of the sleeping and resting areas. The volume containing the spaces to be shared by the family was characterized through slate-stone facades and roof.

The use of recyclable materials and inclusion of energy saving concepts are fully integrated to this work. A wood structure with flax insulation, cellulose, and wood-fiber board was the option selected to obtain a thermal conductivity coefficient of 0.14W/m²K as average value for all the facades. A heating pump with five 35-meter-depth geothermal probes benefits from the earth-generated/stored heat for heating purposes. The great windows placed at the south façade allow the maximum use of natural light, while the 4-meter-hight obtained through vertical shifting to adapt to ground displacement takes advantage of the currents ascending the slope to ventilate interiors in summer.

这栋房子坐落在典型的德国小城镇里，位于慕尼黑市的郊外。这栋房子里的主人是一对已婚夫妇和他们的女儿，他们提出的要求是建造一栋适宜居住的房子，并且在尊重传统风格的同时，体现出独特的现代风格。

在占地面积约为2000平方米的小山坡上，能够领略到小镇的全部景色，设计师便将这栋房子建在这里，取代了原有的农场遗址。严格的城镇条例规定这里只能建造尖屋顶的房屋，这就给委托人所希望的独特设计带来了难题；而且，为了避免周围邻居的不满，他们不得不放弃饱览山谷全景的想法 ，将朝向做了改变。

设计通过一个简单的几何形状，将两个立体空间结合在一起，不仅保留而且还相应地改善了周围住宅的居民观赏山谷的环境。

在休憩的区域，白色永远都是诠释宁静的最佳选择，于是设计师将其设计为板岩石材墙面和屋顶，很好地演绎了可以享受的空间环境。

可回收材料和节约能源的理念，在这个项目中得到了充分的运用。亚麻绝缘的木质纤维板结构被选用作所有墙面的导热材料，其导热平均系数是0.14瓦每平方米开尔文。一个加热泵中带有5个35米深的地热传感器，房屋可以利用这个装置吸收地热为室内供暖；建筑物南侧设有巨大的窗户，可以最大限度地利用阳光；另外，在夏日，可以通过4米高的错层设计来适应落差，还可以利用斜坡处上升的气流进行室内的通风换气。

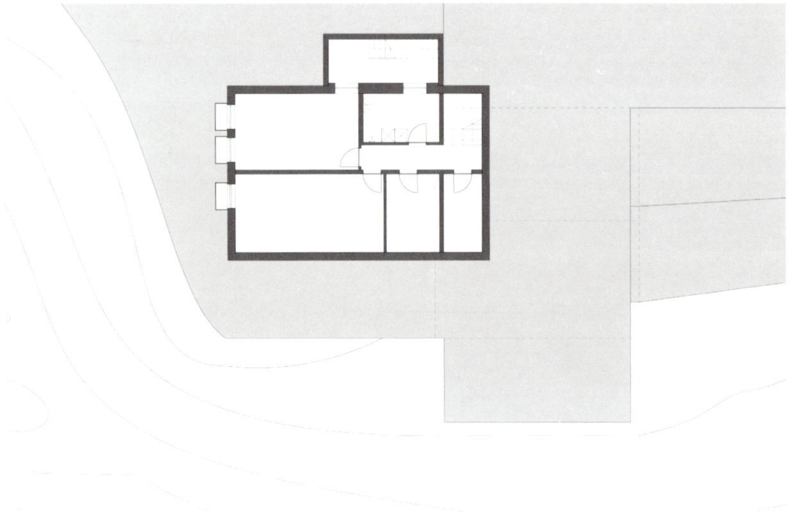

Basement Plan/地下一层平面图

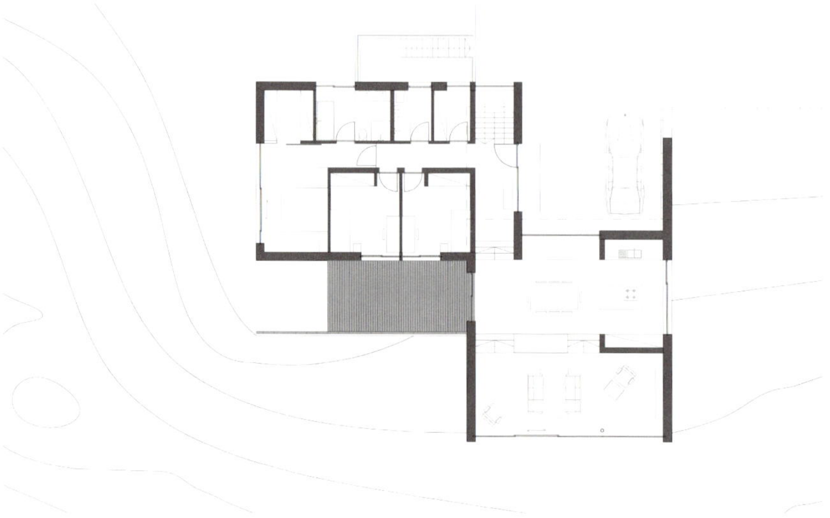

Ground Floor Plan/底层平面图

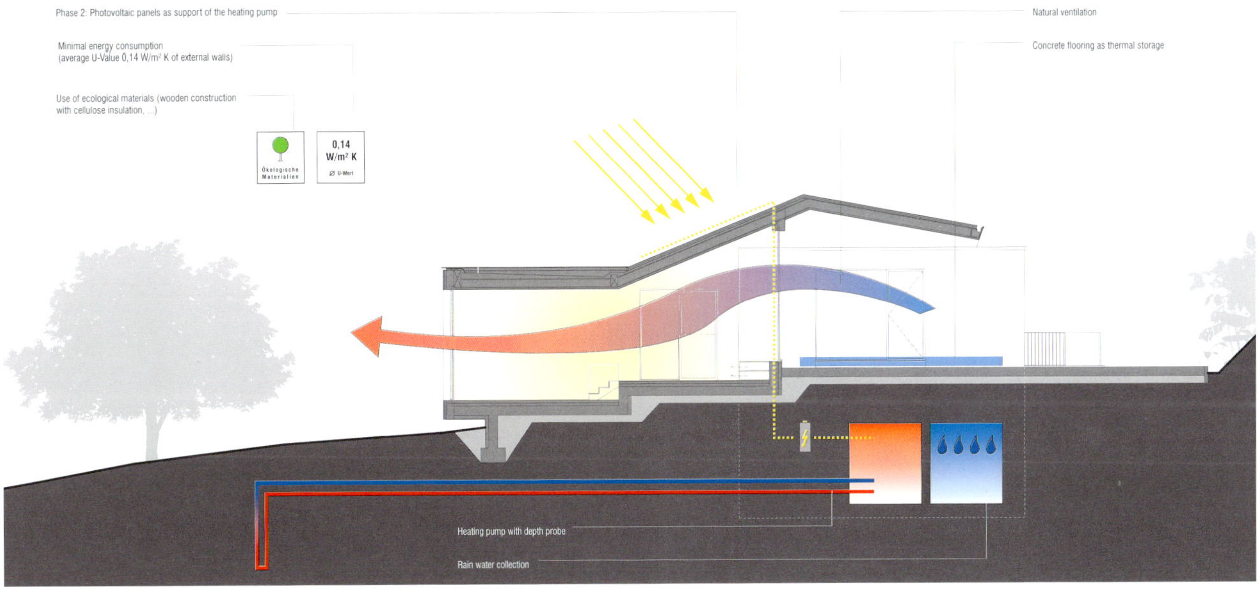
Phase 2: Photovoltaic panels as support of the heating pump
Minimal energy consumption
(average U-Value 0,14 W/m² K of external walls)
Use of ecological materials (wooden construction
with cellulose insulation, ...)
Ökologische Materialien
0,14
W/m² K
Ø U-Wert
Natural ventilation
Concrete flooring as thermal storage
Heating pump with depth probe
Rain water collection

Section/剖面图

House 205

H ARQUITECTES

■ Architecture Design/建筑设计: H ARQUITECTES ■ Project Architect/项目建筑师: David Lorente, Josep Ricart, Xavier Ros, Roger Tudó ■ Location/地点: Catalunya, Spain ■ Area/面积: 132 m^2
■ Photograph/摄影: Starp Estudi, Anna Bonet

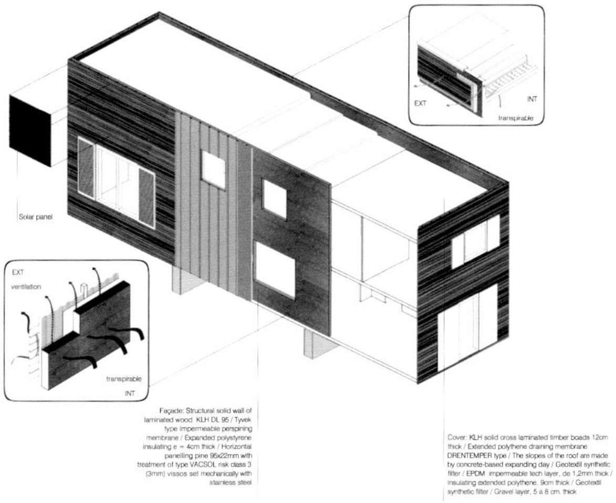

The setting of the project is a plot with steep slopes and a great amount of trees and bushes. The aim is building a house without causing any serious impacts on the land. The house will be built on a natural rocky platform. This platform will also be used as either the exit or the garden of the house.The architects and the property developers have agreed on minimizing land movements in order to build the artificial landscaped platforms. The target is to make good use of the existing land shelves. This will allow us to preserve the natural physiognomy of the wood. The only uneven area will be the path ramp, which will cross the piece of land diagonally. This artificial ramp will communicate the street with the different plot levels.

The large layer of rock and two concrete struts will be used as the foundations of the house. The two struts will lay the foundations of the house on the rock. These two struts will also create a space permanently ventilated between the house and the rocky ground.

This structural system implies a very important diminishment of the weight, raw material, energy and so the CO_2 emissions associated with the foundations and structure of the building. The dry assembly is easy, quick and helps saving great amounts of water. Altogether eases the building and reduces significantly the cost and time investments. The laminated wood is a renewable material that can be dismantled and therefore recycled. This means its life cycle is practically closed. In order to reduce the costs on coating and interior finishing, we have tried to reduce the facings on sight. The installations are hidden but they have just been panelled with gypsum wallboard on those rooms, which distribute the system around the house. Finally, half of the rooms have been left with the structure of its facings on sight.

The facades are ventilated and finished with coniferous wood such as Flanders pine. The openings have also been finished with wooden fitters protected with removable shutters. The transpirability of the facade is guaranteed with a TYBEK panel, which protects the wood. The transpirability of the roof is solved by a draining sheet, which creates a small ventilated room. The final result is a wooden volume set on a large rock and surrounded by forest, which prevents the ecological visual impact.

这项工程设计在一块有着陡峭斜坡的地块上，上面长满了大片树木和灌木丛。其设计的一个重要理念是建造房屋的同时尽量不给当地生态带来负面影响。房屋将建在一块天然的石台上，而石台将同时用作房屋的出口和花园。最低程度地改变土地状态并且建造人工景观平台，在这一点上，建筑设计师和开发商达成了共识。为了能够保持当地林区的地貌，设计师充分利用了现有地块的格局，将仅有的一块不够平坦的地方，修建成一条蜿蜒的小路，斜穿过该地块，最终与街道相连。

一大片岩层和两根混凝土立柱将成为房屋的基石。与此同时，设计师将开辟一块房屋与岩石之间的空间，以保证长期的通风透气。

该房屋结构与其重量、原材料、能源乃至二氧化碳的排放量息息相关。其安装方式很方便快捷，而且还节省了水资源，因此，这样的设计大大减少了房屋的建设成本和时间。所选用的胶合板可拆卸回收，这就意味着该材料可循环使用。为了减少涂料及内部装修的成本，设计师打算尽可能弱化正面的视觉效果。上述所选材料隐藏在室内石膏墙板的控制面板上，系统地分布在这座房屋里，这样，室内一半的房间结构都被遮挡起来。

房屋外部暴露在针叶林中，比如佛兰德斯松树林。房屋窗口也使用了适当的木条，还安装了可移动的百叶窗；TYBEK面板既控制了外部的通道，又可以保护木头；屋顶的通风道安置了排水片，从而形成一个小的通风空间。最终的效果看起来就是一个巨大的木质盒子被安置于岩石上，掩映在灌木丛中，避免了它对该地生态的视觉破坏。

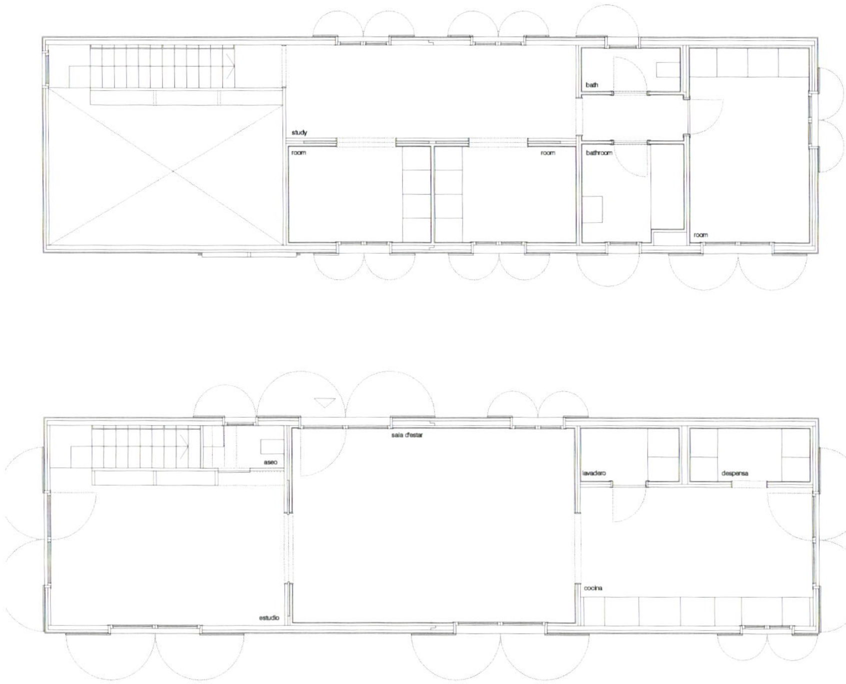

Plan/平面图

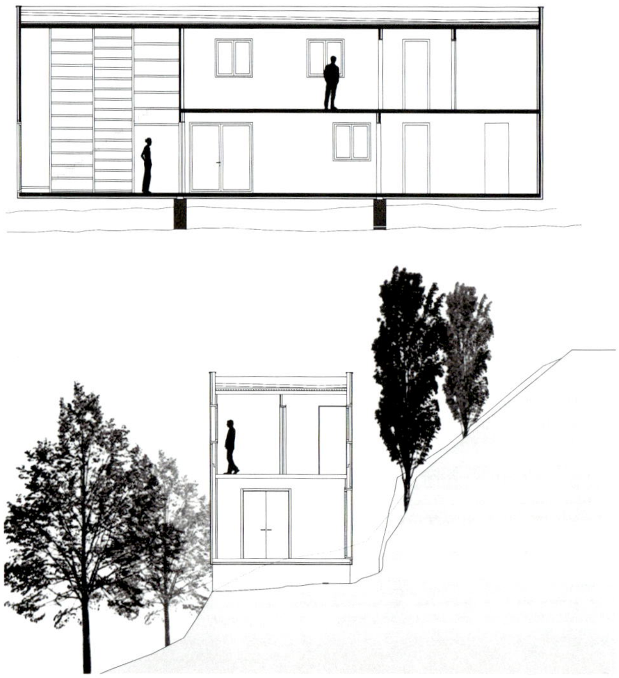

Section/剖面图

Facade:East/东立面图

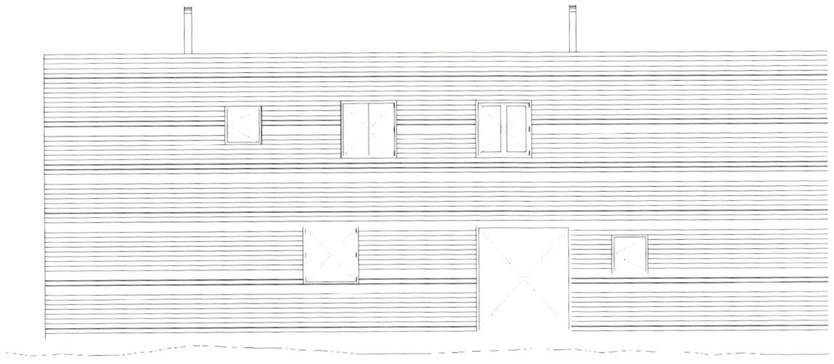

Facade:North/北立面图

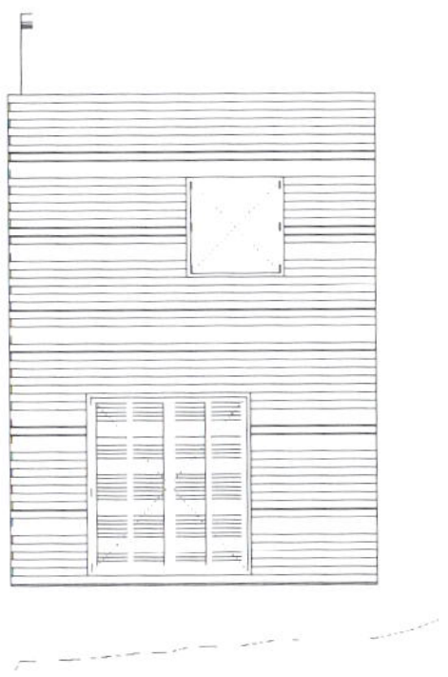

Facade:West/西立面图

Facade:South/南立面图

magnum
COLORFUL

Paradise For Two

NIO architecten

■ Architecture Design/建筑设计: NIO architecten ■ Design Team/设计团队: Maurice Nio, Arek Seredyn ■ Location/地点: Botshol, Netherlands ■ Area/面积: 160 m^2 ■ Completion/建成时间: 2007
■ Photograph/摄影: Hans Pattist

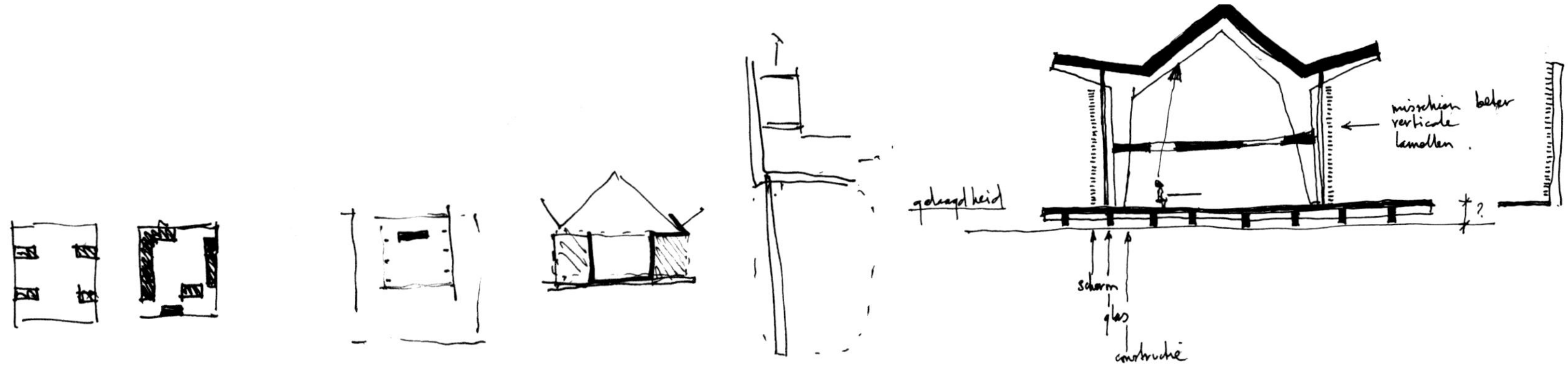

The grass and marshland of the Botshol is a nature reserve and a unique habitat for unusual plants and birds. It is situated - a bit hidden - as an oasis in the midst of the Randstad. It is obvious that in the Botshol cannot be built, but also in the surrounding area of De Ronde Venen there are restrictions to new housing development or partial renewal of existing farms.

Precisely where the austere lay-out of the peat grass area touches the new nature and the whimsical landscape of the Botshol, is a garden of which the inner world does not reveal itself. Almost invisible from outside are two buildings, a farm and an annex, both ready for renewal. In line with the heart of the garden, which has been cultivated and even seems a little heavenly, the renewal was used to intensify the unexpected atmosphere of this location.

The basis of the design are two distinct zinc roofs under which different spaces are given a place in a very relaxed manner. One part is now ready: the annex has been transformed into a pavilion. The zinc roof is cheekily placed diagonally over the square wooden building. The struts of the roof fan out, as if they are trying to imitate the pattern of the nearby Vinkeveense plassen, and the dark stained wooden duckboards are lifted compared to ground level, so that the building seems to float above the garden. The image of the pavilion is new and possibly unexpected, but eventually it is equally natural and graceful as a cormorant in the Botshol.

Botshol地区的草地和湿地是一个自然保护区，也是各种少见的植物和鸟类的栖息地。它地处兰斯塔德中部的绿洲中，有些隐蔽。很显然，这里不仅不能建房，就连De Ronde Venen 地区周边现有农场的重新改造也受到限制。

正是在泥炭色的草地区域上，设计师感受到新鲜又变化无常的Botshol景色，从而将其建成一个鲜为外界认知的花园，从外面几乎看不到两个建筑物——需要重新整修的一个农场和一栋配楼。在花园的核心区域，已经开垦过的上地甚至看起来有一点点神圣感，由此，该次重建为该地增添了意想不到的氛围。

这一设计的基础是两个截然不同的锌质屋顶，设计师将其得体地安放在不同的空间里，其中，配楼部分已成功地变身为小凉亭。设计师将屋顶大胆地安放在斜对面的方形木屋上，屋顶的支柱展开，像是在模仿附近Vinkeveense plassen的造型。地面上褪色的木质踏板，被抬高到差不多一层的高度，这样建筑就仿佛漂浮在花园中。凉亭的画面是全新的，也可能是完全出乎意料的，但在最终，它自然而优雅，仿佛Botshol的一只鸬鹚，令人眼前一亮。

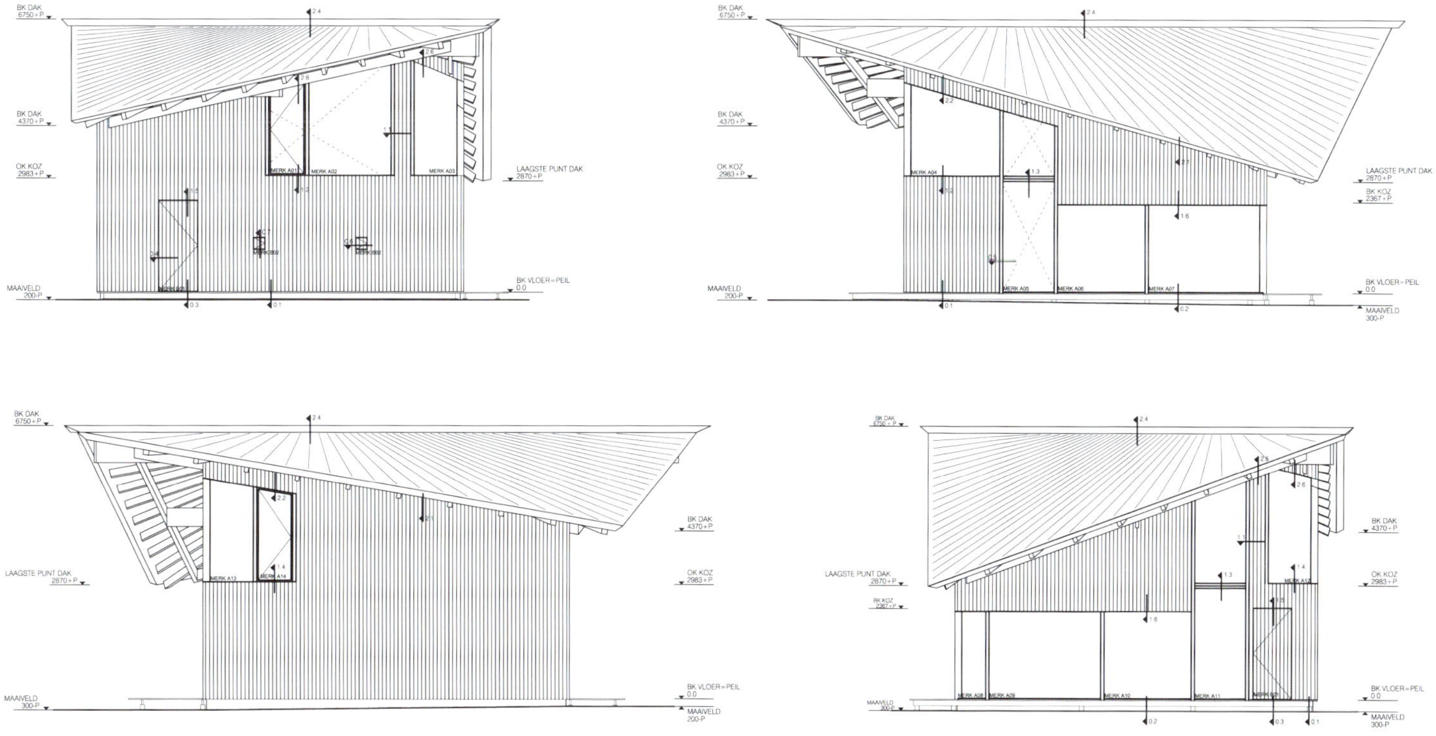
BK DAK
6750 + P
BK DAK
4370 + P
OK KOZ
2983 + P
LAAGSTE PUNT DAK
2870 + P
BK KOZ
2367 + P
BK VLOER = PEIL
0.0
MAAIVELD
200-P
MAAIVELD
300-P

Section/剖面图

Villa Alstrup

C. F. Møller Architects

■ Architecture Design/建筑设计: C. F. Møller Architects ■ Project Architect/项目建筑师: C. F. Møller Architects ■ Location/地点: Hjerting, Esbjerg, Denmark ■ Area/面积: 311 m^2
■ Completion/建成时间: 2009 ■ Photograph/摄影: Helene Hoeyer Mikkelsen

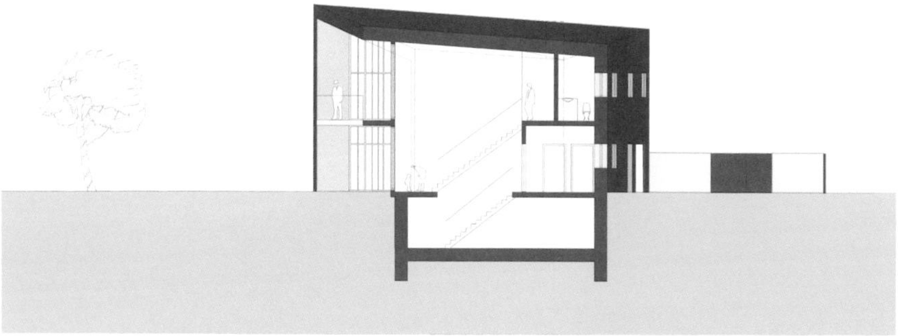

The house on the shore with a view of the Wadden Sea is an energy-plus house, which means that it produces more electricity and heat than it uses. This was achieved without compromising on the exclusive qualities of a large home, including large panoramic sea-views.

The architecture has a clear and simple expression, open and transparent to the sea and more closed and private towards the neighbours. The unusual geometry of the volume is combined with a calm and unpretentious detailing, and a restrained palette of high-quality materials.

The building, which is designed on the basis of the 'passive house' principle, is compact in form, with large windows facing the view to the south-west, which is also the best angle at which to make optimal passive use of the sun's heat. The angle also respects the shoreline protection zone, and out of this derives the triangular floor plan layout, which is one of the buildings characteristics. Another is the sloping roof, which is angled and orientated to optimize the performance of the solar heating and solar cells.

Passive solar heat gain is absorbed and accumulated in the interior concrete walls and floor slabs, while the south-west facing balcony and overhangs shade the facades and control the amount of solar energy in correlation to the seasons. The balcony is designed as a free-standing concrete slab completely eliminating any cold-bridging to the interior.

The house is self-sufficient in electricity and heat through this use of solar and geothermal energy, meaning in effect that it produces more energy than it consumes. By law, the building must still be connected to the grid, but the meters will simply be running backwards in periods.

In the spirit of the project, it is built using efficient pre-fabrication and environmentally-sustainable materials such as paper granules, which is a recycled product, and eco-labelled products. The house is designed to require minimal maintenance; the facade, for example, is clad in fibre cement, to withstand the local Wadden Sea climate. This will help to increase the building's sustainability over time.

这是坐落在瓦登海岸边的一栋产能房。所谓产能房就是表示这栋房子产生的电量和热量比它本身所需的能耗还多。同时，这也是一栋高质量的完美的大房子，它拥有大片完整的海景。

建筑师的设计思路清晰简单：面朝大海的房间要求开放、透明，临近隔壁邻居的方向则需要更加隐私、封闭。房子的整体几何构造与众不同，设计的细节和定制的高品质板面，使房子沉静、低调。

这栋房子的设计源于“被动式节能房屋”原则，外形构造简洁，巨大的落地窗面向西南，这个角度能够对阳光的热量进行最佳的被动式利用，同时这个角度也是根据海岸线保护区而设置的。建筑布局为三角形平面，是该建筑设计的另一个特色。此外，利用屋顶朝向东面的倾斜的角度，建筑最优化地使用太阳能加热系统和太阳能电池板。

自然日照吸收并积累的热量储存在室内的水泥墙面和水泥板中。而朝西南面的阳台外挂于建筑主体之外，这又遮蔽了部分建筑的正面，有利于调控太阳能储量以应对不同季节。

这个建筑物通过使用太阳能和地热能满足对电和热的需求，同时它能够产生大于它自身能耗的能量。依据法律规定，建筑物必须与电网连接，因此，电表在使用期间是逆行运转的。

这个项目的精髓，就在于使用了高效的预制件和环保的材料，如纸张、颗粒等再生产品和可回收产品。这栋房子的设计要求，在于建筑材料使用量的最小化。另外，在墙面上使用纤维水泥来应对瓦登海当地的气候，这将有助于日后增加建筑物的持久性。

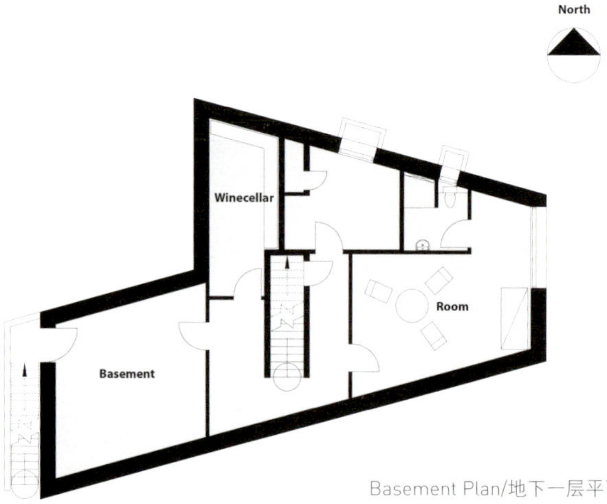

Basement Plan/地下一层平面图

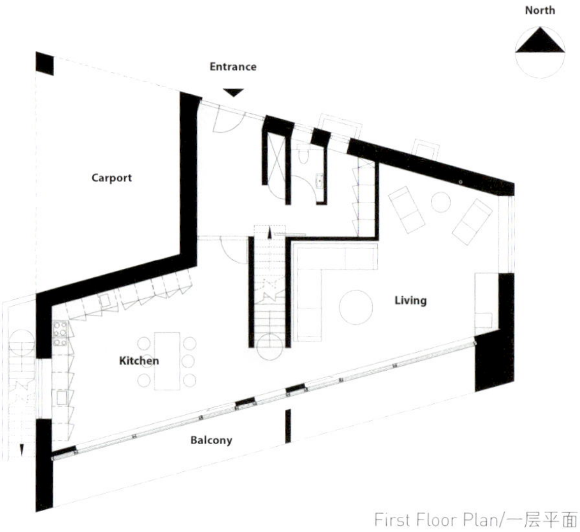

First Floor Plan/一层平面图

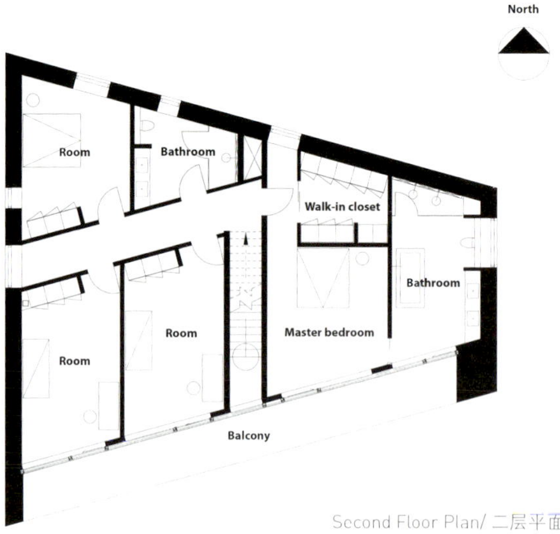

Second Floor Plan/ 二层平面图

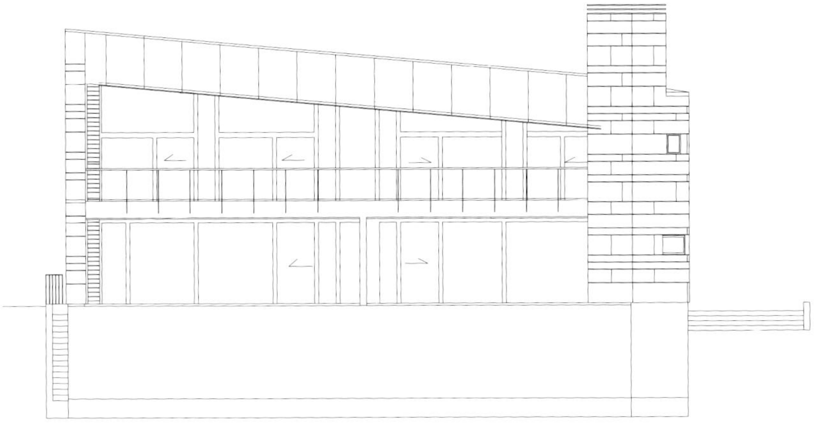

Facade: West/西立面图

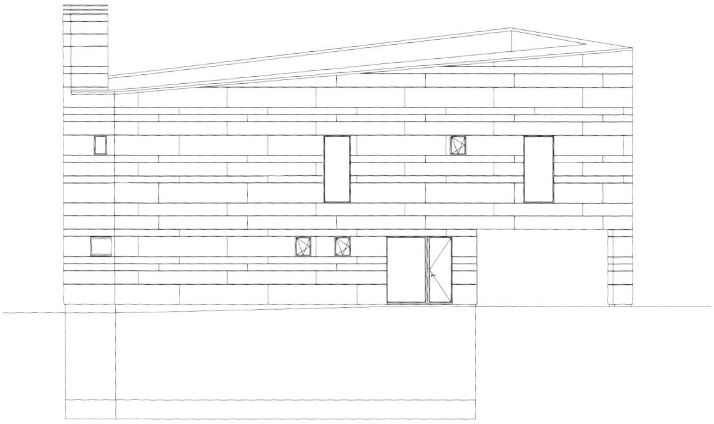

Facade: East/东立面图

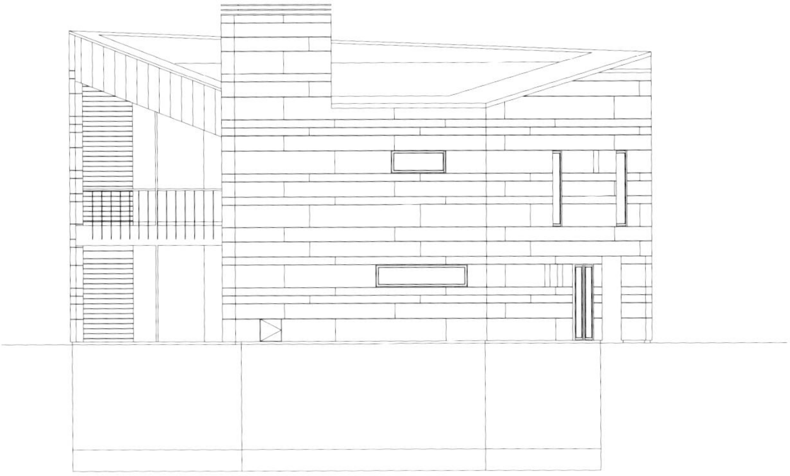

Facade: South/南立面图

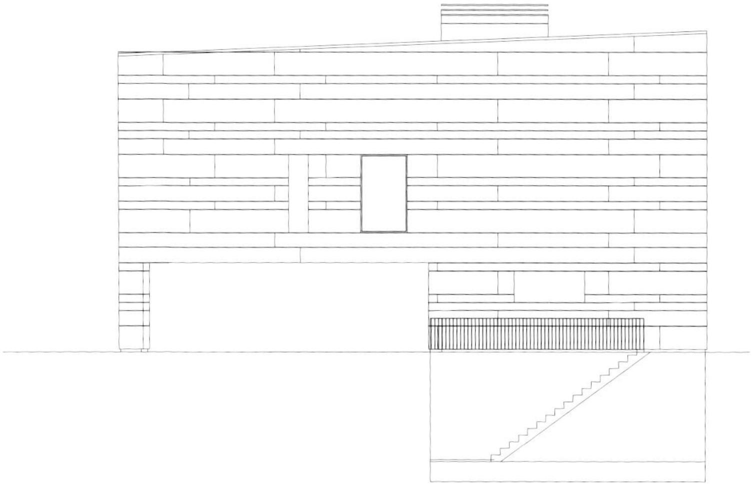

Facade: North/北立面图

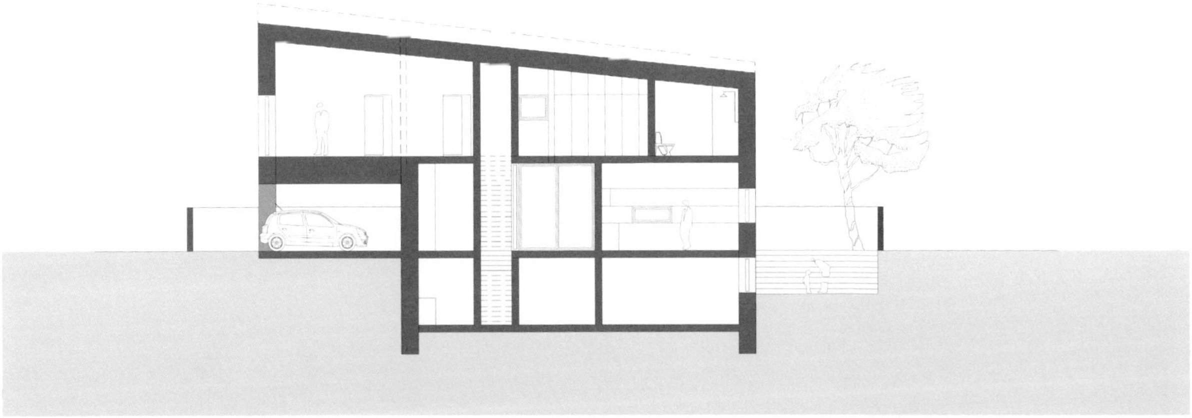

Section/剖面图

House R in H - Wedge House

Matthias R Schmalohr DI Architect BDA DWB

■ Architecture Design/建筑设计: Matthias R Schmalohr DI Architect BDA DWB ■ Project Architect/项目建筑师: Matthias R Schmalohr ■ Location/地点: NRW, Germany ■ Area/面积: site 900 m²/building 224 m²/floor 266 m² ■ Completion/建成时间: 2008 ■ Photograph/摄影: Klaus Dieter Weiss

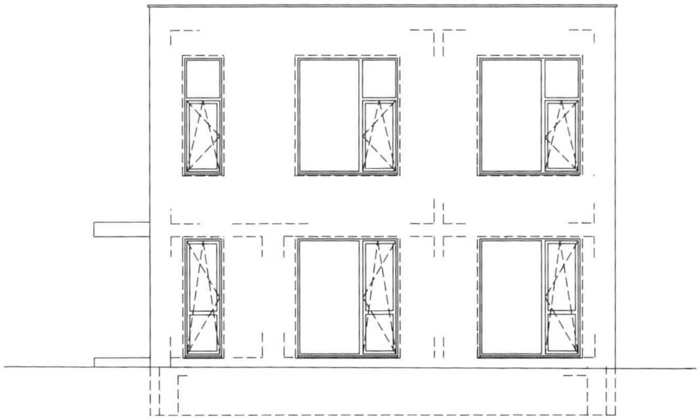

Scenic uniqueness all to often suffers from being surrounded by a very ordinary variety of architecture. In general the detached house targets at 'uniqueness', too – though a very individual, or rather commercially shaped uniqueness. This is also true for the mountain range of the Wiehengebirge, located next to 150 km of the northern German lowlands. The house is situated in the immediate neighbourhood of the Wiehengebirge and upgraded by the Midland Canal with its vegetated banks. The surplus of uniqueness without any consent does not lead to an appealing, specific situation, shaped by scenery and architecture. It leads to an undesigned chaos of form, colour and natural as well as urbanistic space.

At this place there is introduced an additional architectural element, but his design demonstrates a halt, a standstill. It contrasts the jolly airiness of an undirected collage of everything with architecturally established urbanistic targets. A white signal for the possibilities of unity in variety, the argument for an essentially different, better treatment of the scenery. Not by raising the educational voice, but with the great space and freedom of modern living. For authorisation reasons the roof had to be pitched. But it is not a red, yellow, black or blue saddleback roof. The inside follows the outside form and thus creates an island of silence amid chaos. The occupants immerse themselves in the treasure of an undreamt of quality of living conditions.

独特的景致常常被一些平庸的建筑包围。通常情况下，独立式的房屋也会在“独特”上大做文章——甚至有时会建造出那种非常个性或商业气息很浓的奇特房屋。而这种情况在距离德国低地北部150千米处的维恩山脉附近，也同样存在。该建筑坐落在距离维恩山脉最近的地方，这里的地势被两岸丰茂的中部运河抬高。这些未经许可而私自大量建造的多余建筑，没能与景色融为一体，相反，却使该地呈现出带有多种形态和色彩，这些或自然或城市化的建筑堆积在一起，显得杂乱无章。

这里要引入一种额外的建筑元素，它将这些无序拼接的设计气息与确定化的城市设计目标相对比。这是一场本质上不同的争论，是为了改善周围环境与景色，不是要提高充满着教育意味的呼声，而是要对现代生活方式给予足够的空间和自由。经委托方授权，该建筑物的顶部只能是尖耸的；屋顶也不能是红色、黄色、黑色或者蓝色的鞍形屋顶；建筑物的内部结构要与外部结构相一致。这样　来，室内空间就得以闹中取静，而这里的居民则会沉浸在梦寐以求的自由生活中。

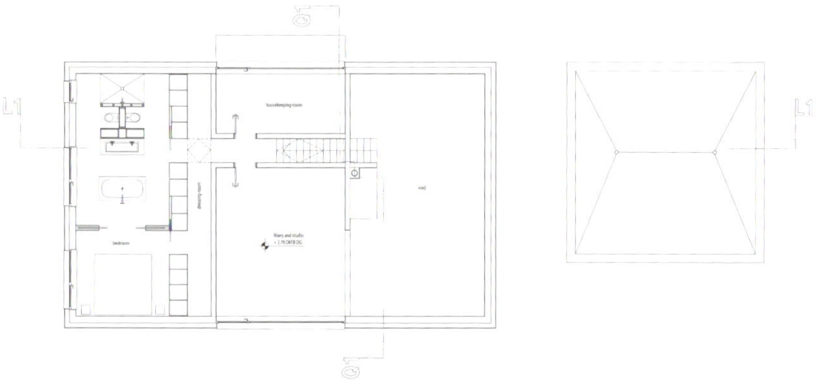

Second Floor Plan/二层平面图

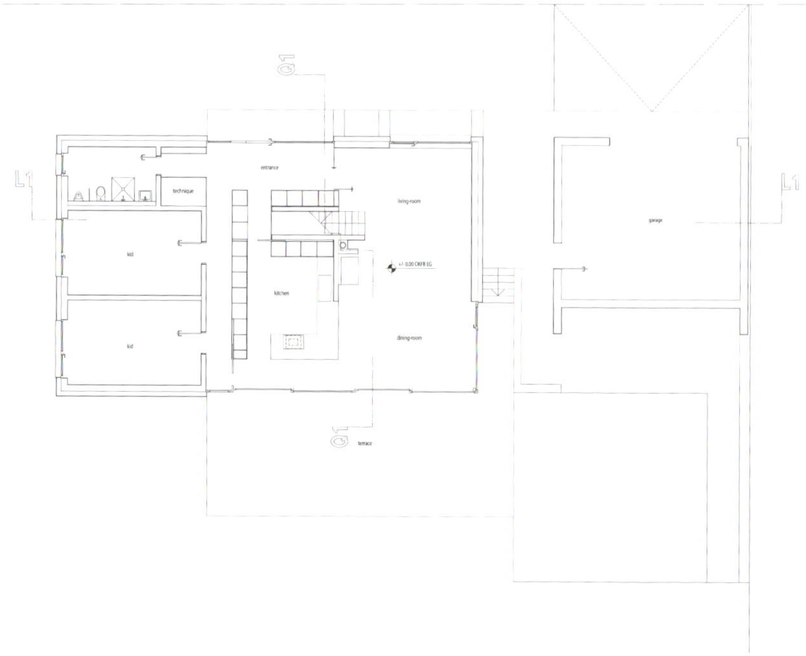

First Floor Plan/一层平面图

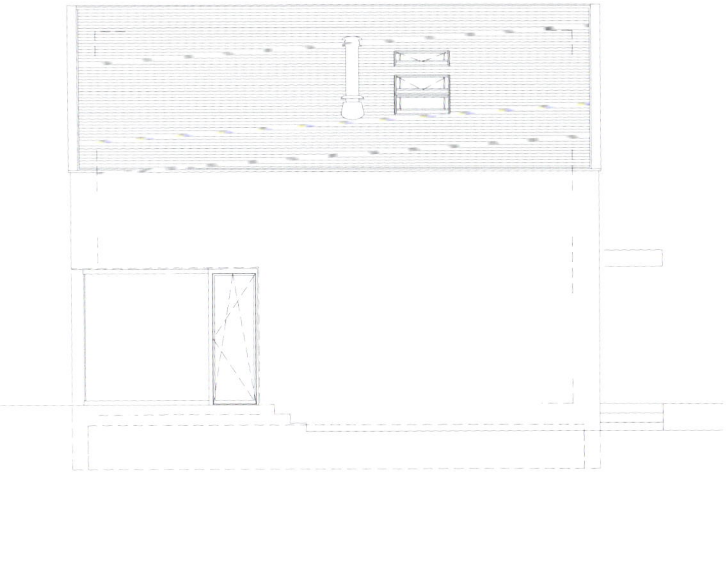

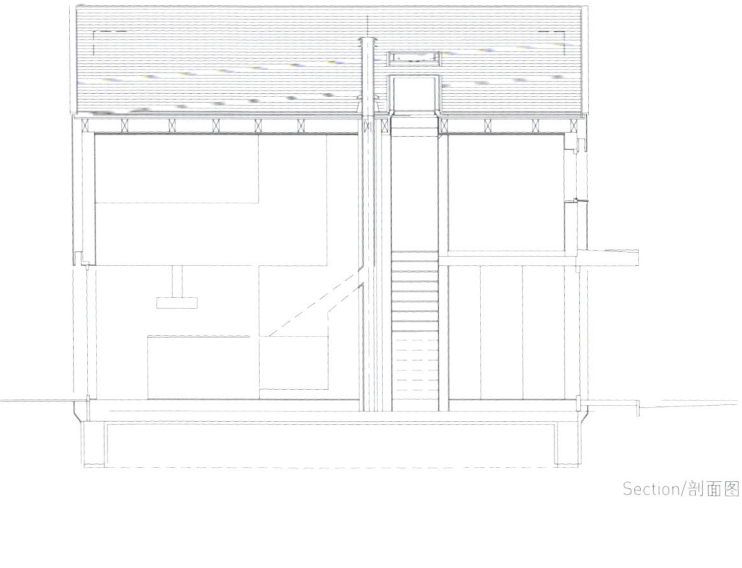

Section/剖面图

Facade/立面图

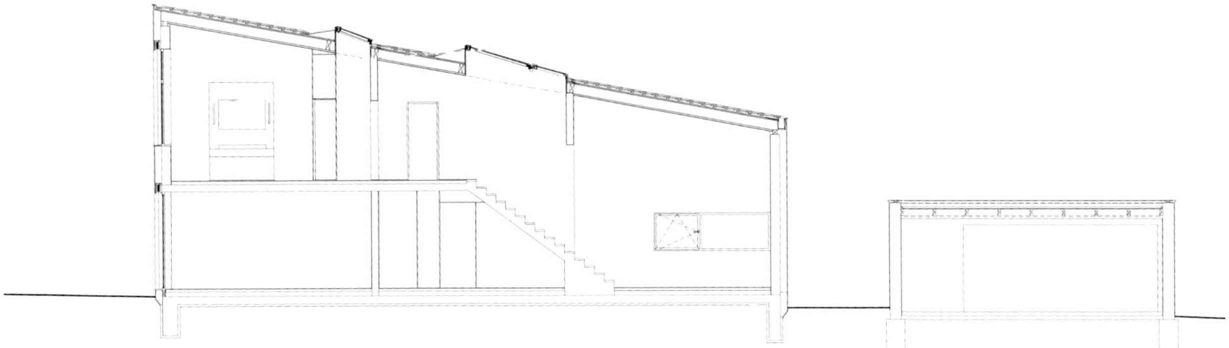

Section/剖面图

House Baetens

JagerJanssen architects BNA

■ Architecture Design /建筑设计: JagerJanssen architects BNA ■ Design Team/设计团队: Alex Jager MSc, Rogier Janssen MSc, Rob Bergervoet MSc ■ Location/地点: Netherlands
■ Area/面积: 220 m^2 ■ Photograph/摄影: René de Wit

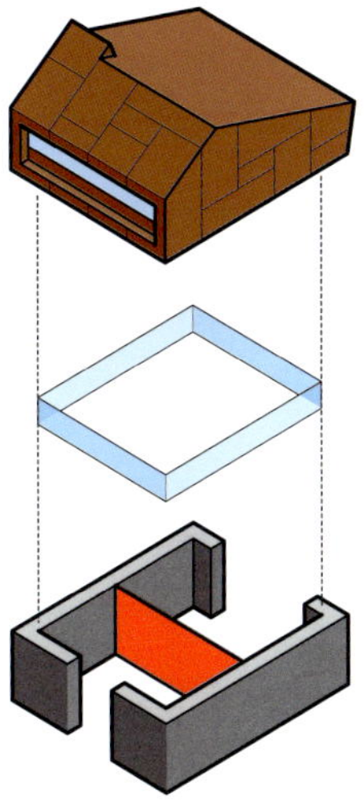

Anticipating the arrival of their first child, private constituents Sandor and Judith were looking out for a new home. Together with architect-to-be Rob Bergervoet, JagerJanssen designed the solitary dwelling on the 618 sq. meter plot, part of urban project Westeraam in Elst, The Netherlands.

The constituents aspired a modern starters villa, to house their young family in a spatial fashion. Daily activities are concentrated in a large open space on ground level, enabling unobstructed view on playing infants. In addition, interior office space was required. The private functions are located on the first floor.

The dwelling consists of two layers. The texture of the lower layer is heavy and earthly. Dark bricks with emphasized horizontal seams give the base an abstract and rock-like feel. The base carries the – in multiple respects – lighter top layer, a wooden monolith. The seams in the Trespa HPL cladding façade suggest that the volume consists of massive blocks.

The separation is emphasized by a separating horizontal glass strip. A soft effect is reached by using wobbled double layered translucent plastic plates in the front and rear of the upper floor façade. The front façade looks more formal due to the central placement of the window frame. In the lower left the flower casing turns around the corner, introducing the less rigid side façade, which culminates in far more transparent rear façade.

The lower floor is constructed around an axis leading from front- to terrace door. Perpendicular to this axis a wall spanning from side to side marks the border between living- and service spaces. The living space is a synoptic space which also contains the separate cooking unit. The part of the wall that borders to the kitchen contains the built in cooking equipment.

From the landing of the straight staircase, there is a nice view at the triangular window. On the upper floor the infrastructural axis is oriented in the other direction, thus creating an interesting spatialplay.

当时，桑德尔和朱迪思这对恩爱的夫妻为迎接他们第一个宝宝的降生，正寻求一个合适的新家。于是，他们与设计师罗伯・博格沃特，耶格・朱瑟一起，在占地面积为618平方米的地块上，设计了该建筑，使其成为荷兰Elst地区Westeraam城市规划项目的一部分。

这对年轻的夫妇希望能找到一所现代化别墅，作为他们组建家庭的开始，从此在这栋大房子里过上时尚的生活。日常的生活主要集中在一层宽大的开放空间里，这样能够在相对宽敞的环境中照看玩耍的婴儿。另外，设计师还设计了在家办公的场所，类似于这种私人化的生活设施都集中放置在别墅的一层。

从建筑的质感讲，底层沉稳、传统。一层层密实平行的青砖使该建筑的基础看起来既抽象，又透出砖石古朴的气息。而在底层之上，设计展现了建筑多元化的一面：较为轻型的木质上层结构。特莱仕高温层压制密板材的运用，也就决定了要使用大块材料作为建筑表面的材料。

设计师将一个水平的玻璃分隔带安放在两层之间，强化了分割感。在楼上，前后外墙使用的是磨砂玻璃，这种半透明的形式在一定程度上弱化了该建筑的风格。由于正面的窗框居于整体的中央，因此从正面看，建筑显得格外庄重。在一层左侧的拐角，种植着花草，点缀着僵硬的墙面。

别墅的一层是以房前到阳台门为轴而构建的。与该轴线垂直的一面墙贯穿建筑两侧，将室内的生活区和娱乐区分开。生活区是一个温馨的空间，包括了独立的厨房，靠近厨房的墙面上还安装了内置式厨具。

走上笔直的台阶，来到一个全新的空间，还可以透过三角形的窗户，领略室外的美景。楼上基础设施的轴心朝向另一个方向，这样就产生了一个有趣的空间转换。

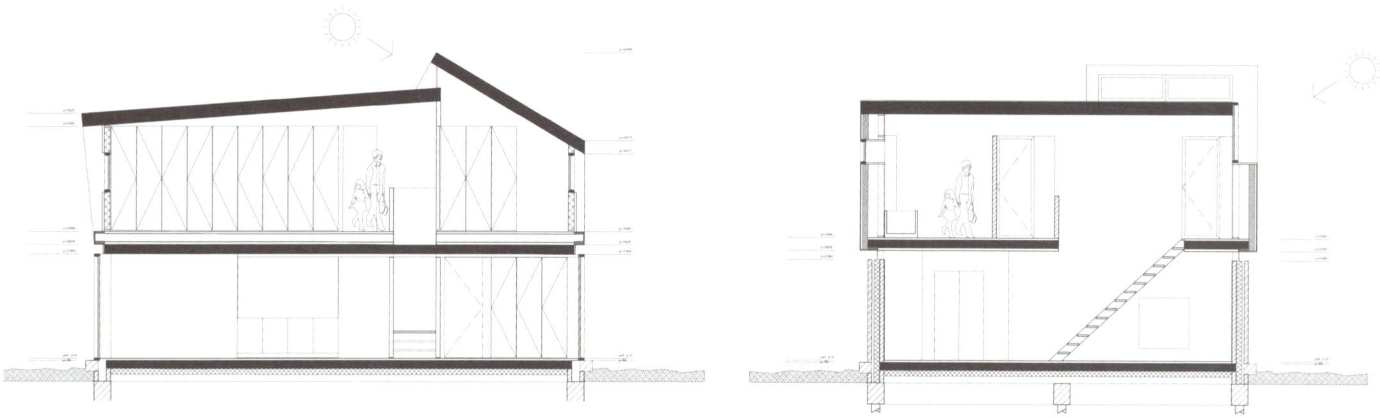

Section/剖面图

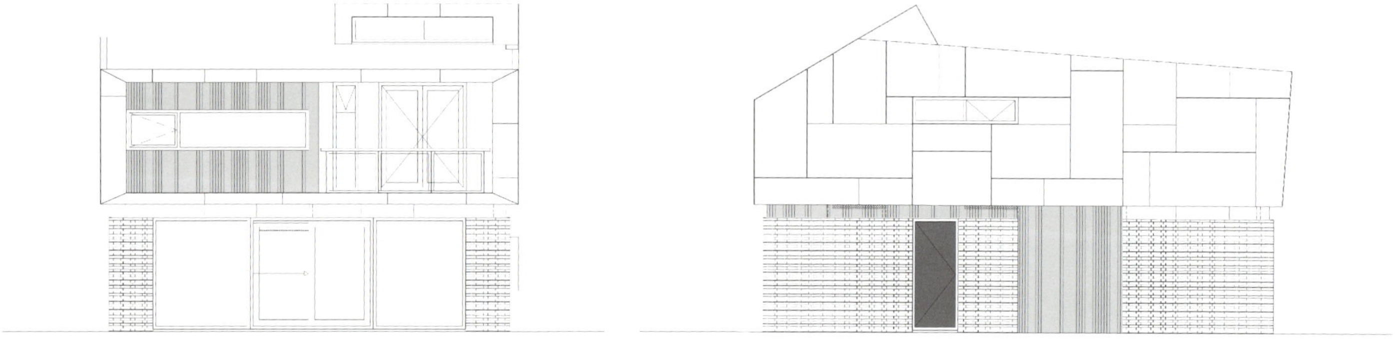

Facade/立面图

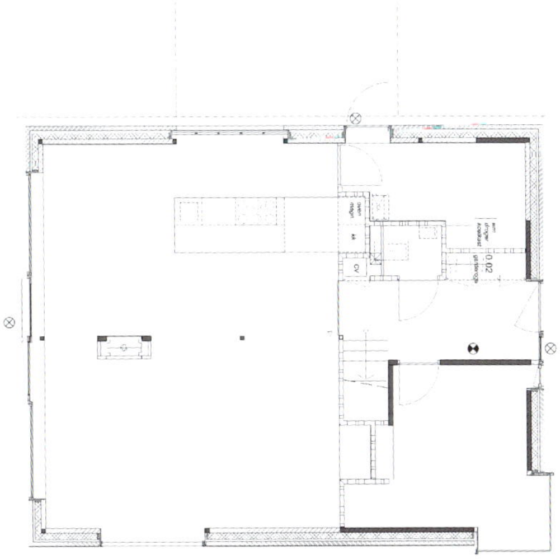

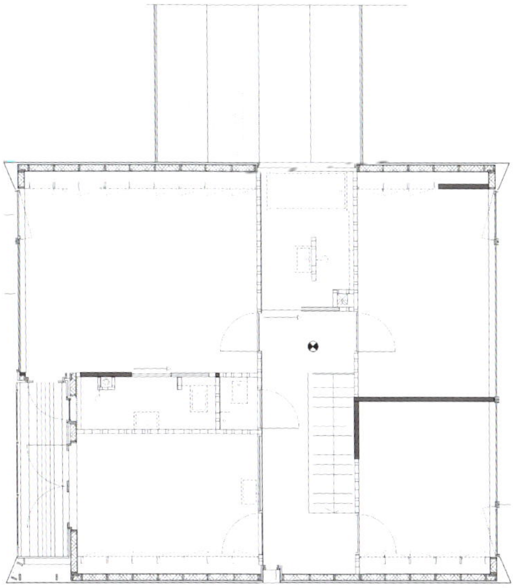

Plan/平面图

House HanenDick

JagerJanssen architects BNA

■ Architecture Design /建筑设计: JagerJanssen architects BNA ■ Design Team/设计团队: Alex Jager MSc, Rogier Janssen MSc, Marijn Boterman ■ Location/地点: Netherlands ■ Area/面积: 220 m^2
■ Photograph/摄影: Rob 't Hart

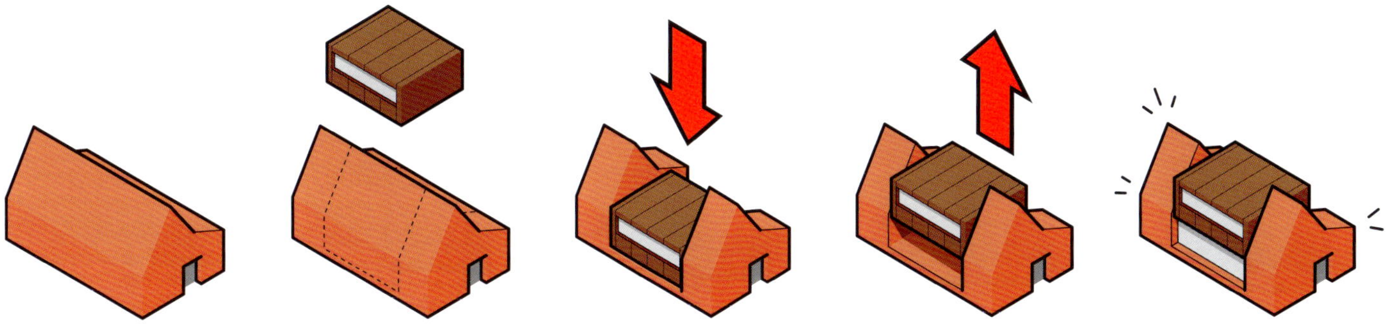

Early 2004, the private commissioners Dick and Hanneke bought a splendid 1200m2 lot adjacent to the Kosterijweg in the centre of Eelde, The Netherlands. The original building on the plot was ancient little farmhouse in deplorable state, which had to be demolished.

Requirement and desires The location as one of the oldest streets in centre of Eelde was the main interest of Hanneke and Dick. The old trees on the premises were to be left untouched.

Dick insisted on a single free standing volume on the site, allowing traffic around the building. A garage was not part of the requirements, a bicycle storage was. The residence should be light and harmonious. Hanneke teaches piano, this required a separate infrastructure for students. In addition she wished the piano to be interchangeable between teaching room and living, Dick however wanted to preserve privacy during lessons. Due to Hanneke's and Dick's busy social life, the dwelling should be hospitable: a official front door for salesman and a informal entrance for family and friends. Also, enjoying the morning sun must be made possible. Both Hanneke and Dick were attached to sufficient storage space, although Dick explicitly did not want an attic. On the first floor not less than 4 bedrooms had to be realised, in case all three sons were to pass by. At last, Hanneke would like to have a shared balcony for the bedrooms.

Modus operandi From the beginning, JagerJanssen aimed at preserving the character of the village street. It should be a modern dwelling with respect for the past. By maintaining the original footprint and the typical fold in the roof, JagerJanssen searched for a connection. In the materialisation a link with the neighbouring farmhouses, church, museum and the other dwellings of the Kosterijweg would be established.

The Result The result is a brick volume, penetrated by a box which seems to have fallen into the main mass. The brick house is kept pure and abstract by continuing the bricks onto the roof.

早在2004年，该项目的委托方迪克和汉尼卡就在荷兰Eelde地区的中心、邻近Kosterijweg的地方购置了一处1 200平方米的地皮。该地的原有建筑是一个古老的小农舍，由于其结构已经破败不堪，因此只好将其彻底拆毁并重建。

要求与期待 该建筑位于Eelde地区中心最古老的街道，也正是出于这个原因，迪克和汉尼卡才对这里情有独钟。原址上的老树被要求保留下来。

迪克坚持在此地建造独立的建筑，这样周围的道路依旧可以通行。在这里不需要为汽车专设一个车库，只要有停放自行车的地方就好，住在这儿的主人应该具备环保和谐的意识。汉尼卡是钢琴老师，因此房子里需要有独立授课的基础设施，她还希望钢琴能够在教室和客厅间交替摆放，同时，迪克希望授课时依旧可以保持私密性。考虑到两位忙碌的生活，居住环境要设计得非常宜人：正式的阳台前厅可以接待来访的陌生人，还要有一个供家人朋友使用的日常通道，同时，房间内每天早上还能够接受到阳光的照射；迪克不想在别墅中增加阁楼，但他们想要足够的储物空间；一层至少得有四间卧室，以免他们的三个儿子同时到来；最后，汉尼卡想有一个公共阳台，可供多个卧室使用。

实际操作 起初，JagerJanssen公司的设计核心是使建筑保留与街道一致的风格，将其建成一栋带有怀旧感的现代房屋。他们在保存建筑原有遗迹和典型折叠式屋顶的基础上，寻找到一种联系。而这种联系是通过使用特定的建筑材料，与附近的农场、教堂、博物馆和其他Kosterijweg的居民区建立起来的。

最终印象 建筑特点以砖房为主框架，在中间插入了一个像盒子一样的玻璃屋。建筑的屋顶延续了主框架的砖石材质，保持了建筑的一致性。

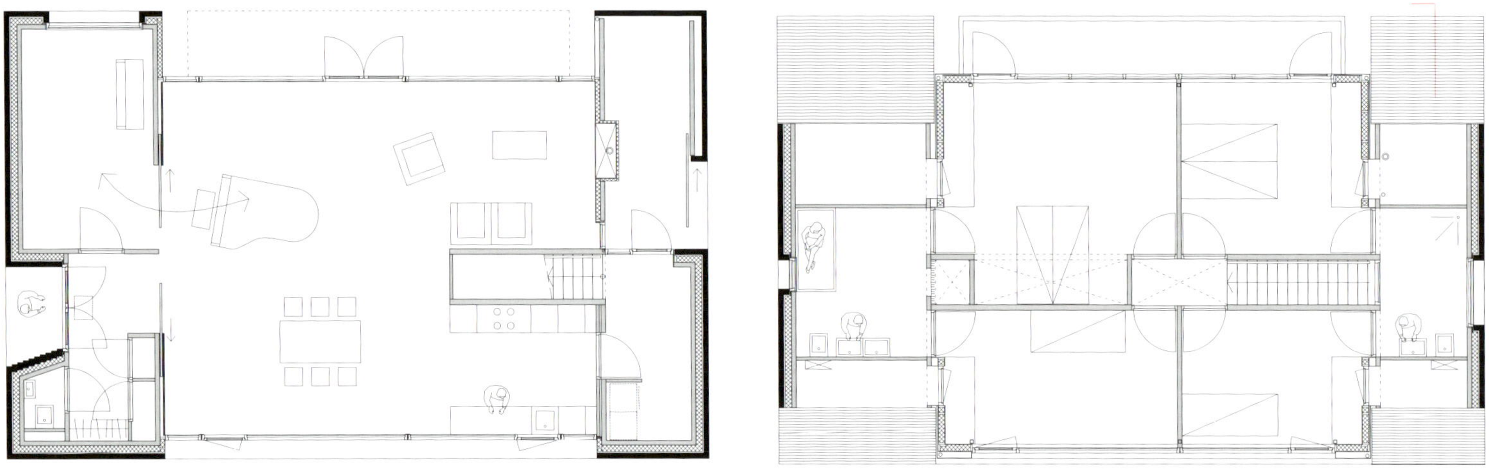

Plan/平面图

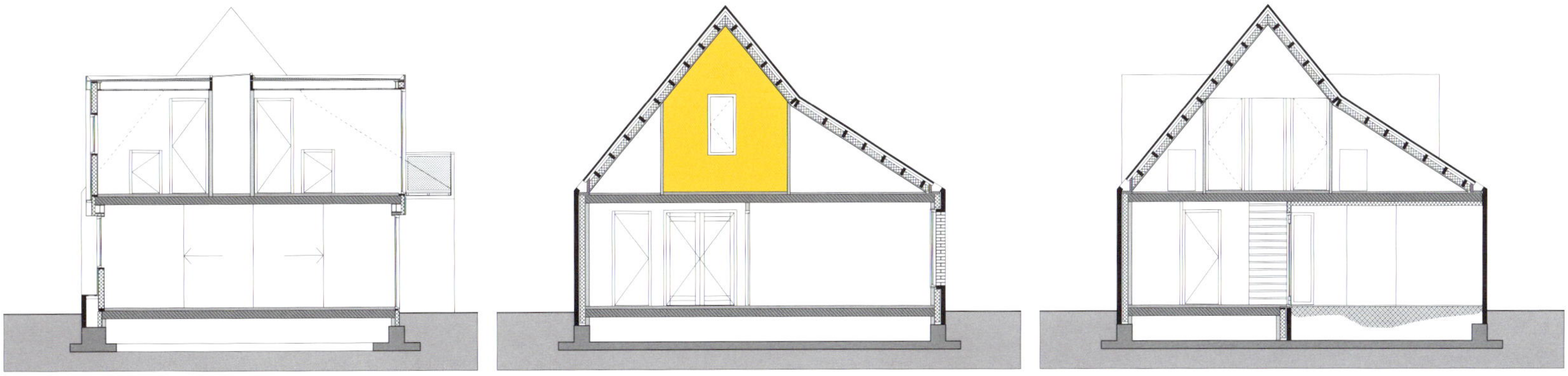

Section/剖面图

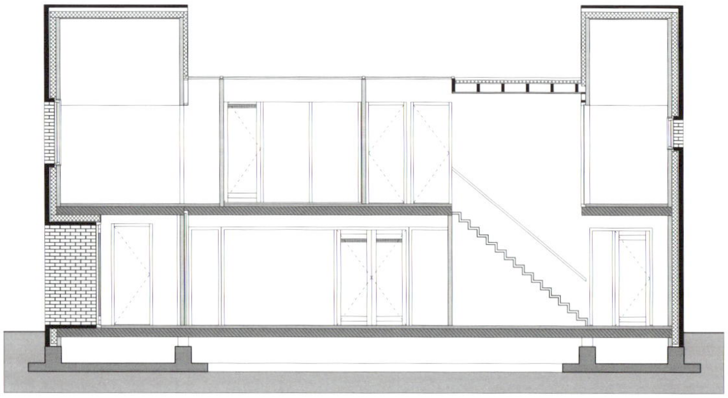

Section/剖面图

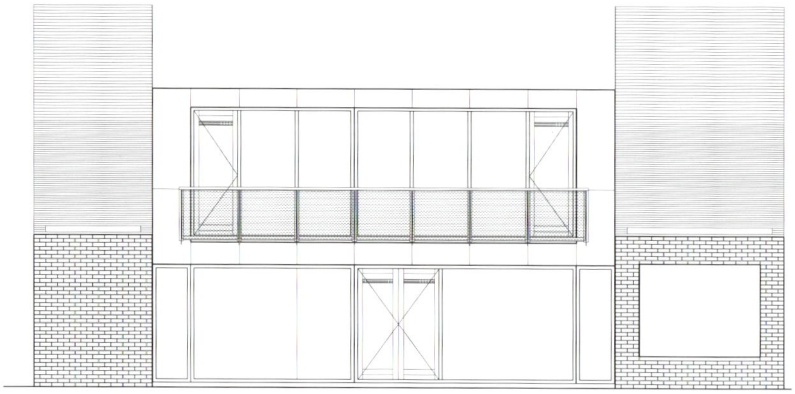

Facade:West/西立面图

Archipelago House

Tham & Videgård Arkitekter

■ Architecture Design/建筑设计: Tham & Videgård Arkitekter ■ Project architect /建筑师: Martin Videgård, Bolle Tham ■ Loation/地点: Stockholm archipelago, Sweden
■ Area/面积:130 m^2 +terrace ■ Photograph/摄影: Åke E:son Lindman

The starting point is the direct relation to the dramatic archipelago landscape with the objective to offer within a simple frame - a platform - several diverse readings of the relation space-nature.

Conceived as a light weight construction in wood and glass, this summerhouse is built in the outer Stockholm archipelago. The horizontal character of the black stained exterior relates to the verticals of tall grown pines and the mirrored views of the Baltic Sea.

The geometry of the plan is generated by the specifics of the site, the house being fitted into the flat surface between two mountain rocks and turning simultaneously towards the sun in south and frontally towards the sea in west.

With the small rooms placed in the back, the social areas of the house stand out as an open platform crisscrossed by sliding glass. The zig-zagged layout also creates a series of outside places sheltered from the strong winds.

该建筑设计的理念源于这个群岛的风景，这个群岛给人们留下了非常深刻的印象。因此设计师希望依托这里的美景修建一个平台，再准备一些不同种类的关于空间与自然的“读物”。

这个凉亭建在Stockholm archipelago项目之外，设计选用的建筑材料是重量较轻的木材和玻璃，将褪色的水平黑框、高大粗直的松树与映着波罗的海风景的玻璃窗搭配在一起。

几何图形的设计是根据建筑物的具体位置而定的，设计将房屋建在两个山石礁和岔路口之间平坦的地方。与此同时，房子的南面可享用充足的阳光，而房子西侧的海景则可以让人大饱眼福。

较小的卧室全部设计在房子的背面。活动区域被布置在一个开放的平台上，还配有纵横交错的滑动玻璃窗。曲折的布局同样也创造了一系列能够遮风避雨的户外空间。

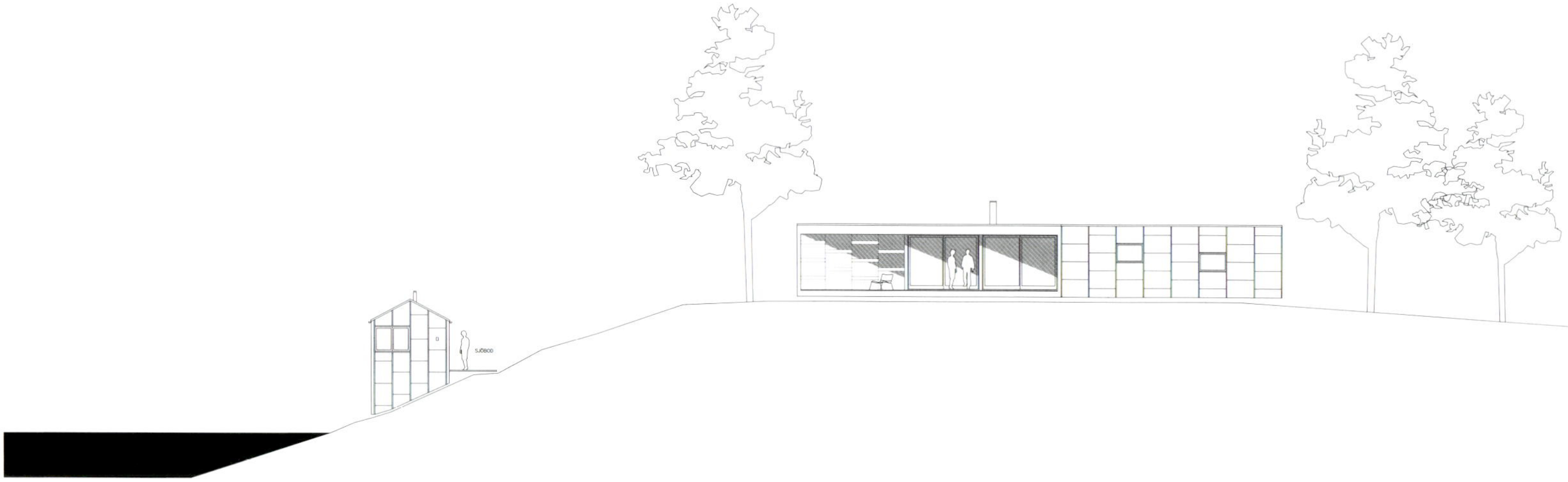

Section/剖面图

1. ENTRANCE
2. KITCHEN
3. LIVING ROOM
4. MASTER BEDROOM
5. CHILDRENS ROOM
6. BATHROOM / LAUNDRY
7. STUDIO
8. GUEST ROOM
9. TERRACE / WOODEN TRELLIS

Plan/平面图

Shell

ARTechnic architects

■ Architecture Design /建筑设计: ARTechnic architects ■ Project Architect/项目建筑师: Kotaro Ide, Moriyuki Fujihara, Ruri Mitsuyasu, Takashi Mototani, Kenyu Fujii ■ Location/地点: Nagano, Japan
■ Area/面积: site 1,711.03 m^2 / building 274.38 m^2 / total floor 329.65 m^2 ■ Completion/建成时间: 2008 ■ Photograph/摄影: Nacasa & Partners Inc.

Being in sync with nature isn't about yielding to nature – it's about coexistence. The existence of the structure depends on its power to endure nature. By isolating living space from the wilderness, and upgrading its quality as a shelter, the house will be protected from nature and will provide a comfortable environment. With this, the house will be taken care of and used frequently and continuously.

The central control system enables all mechanical and electrical equipments to be managed by three buttons. In addition, the biometrics lockage and security system will reduce anxiety and stress over house safety management.

The plan was to build the villa around the big fir tree as the center of the site, with a row of pine trees as the main view. Initially, we had planned to build a shell structure with three dimensionally curved surfaces, and the C shaped section was to surround the fir tree and the plan of the building resembled the letter J In addition, certain parts were planned to hold double volume space. Yet, going over the budget, construction method and finish, the plan was revised down to a shell structure of two dimensionally curved surfaces. The J shaped structure is constructed by two different size oval cylindrical masses cut with curves. The straight part of J, a smaller mass connects to a curved part to J, a larger mass. The top of the oval shaped building wall thickens by 350mm and its width continuously increases up to 750mm at both sides to meet the structural requirements. The free-curved lines appear on the edge, and the three dimensionally curved surface with a twist partly appears on the cut surfaces. However the entire structure was composed by two dimensionally curved surfaces. The floor is built 1400mm above the ground, with the lower half of the shell structure protruding greatly towards the outside, supporting the terrace of the same height. All air and exhaust outlets are installed beneath the sash, letting air run outside through the terrace louver. In addition, by devising unfixed windows, we tried to maximize natural ventilation (we haven't arranged air conditioning in general parts). While at a glance, the oval shaped cylinder space might appear as wasteful use of space, the functional use of space is maximized by the installation of furniture in the lower half of the oval cylinder.

Considering the often short and intermittent use of villas, the expanded hard urethane form has been installed to cover 60 mm of the interior wall of the oval cylinder, which will reduce thermal capacity.

建筑物要顺应自然，并不是向自然界屈服，而是要与之和谐共存。该项目建筑地处气温低、湿度高的恶劣气候区，因此，传统结构的房屋不同程度地遭受到自然的侵蚀。如何做到与自然的和谐共生而不失舒适性，成为该设计的重点。

中央控制系统全部机械化、电气化，只需三个按键即可完成，生物识别增强了安生性。

本案最初计划以巨株冷杉树为核心，以联排松树为景观。建筑为三维曲面壳结构，“C”形截面绕冷杉树而建，同时建筑呈字母“J”状，部分区域设计高度为两层。但由于预算紧，建造方案复杂，完工时间长，设计方案只好由三维曲面改成平面，曲面壳体结构仍旧保持。“J”形结构由两个不同大小的椭圆形圆柱曲线切割而成。“J”形直线部分比例小，曲线部分比例大。壁顶厚度350毫米，壁的厚度由核心至边缘逐渐加深至750毫米，以满足结构要求。边缘线条自由弯曲，三维曲面切割部分扭转叠加。地板高丁地面1400毫米。壳结构凸起部分向外伸张，从而对相同高度露台起到支撑作用。气口位于窗框下方，通风系统与阳台和谐共存，不固定窗口的设立实现了自然通风。柱体、椭圆形空间的设计看起来似乎浪费了一定的空间，实际上却使椭圆柱体下部的家具摆放达到了最大化。

考虑到别墅使用的短期性、间歇性，椭圆柱体空间内墙加大了硬聚氨酯材料的使用，其厚度高达60毫米，从而更好地起到防潮、隔热的作用。

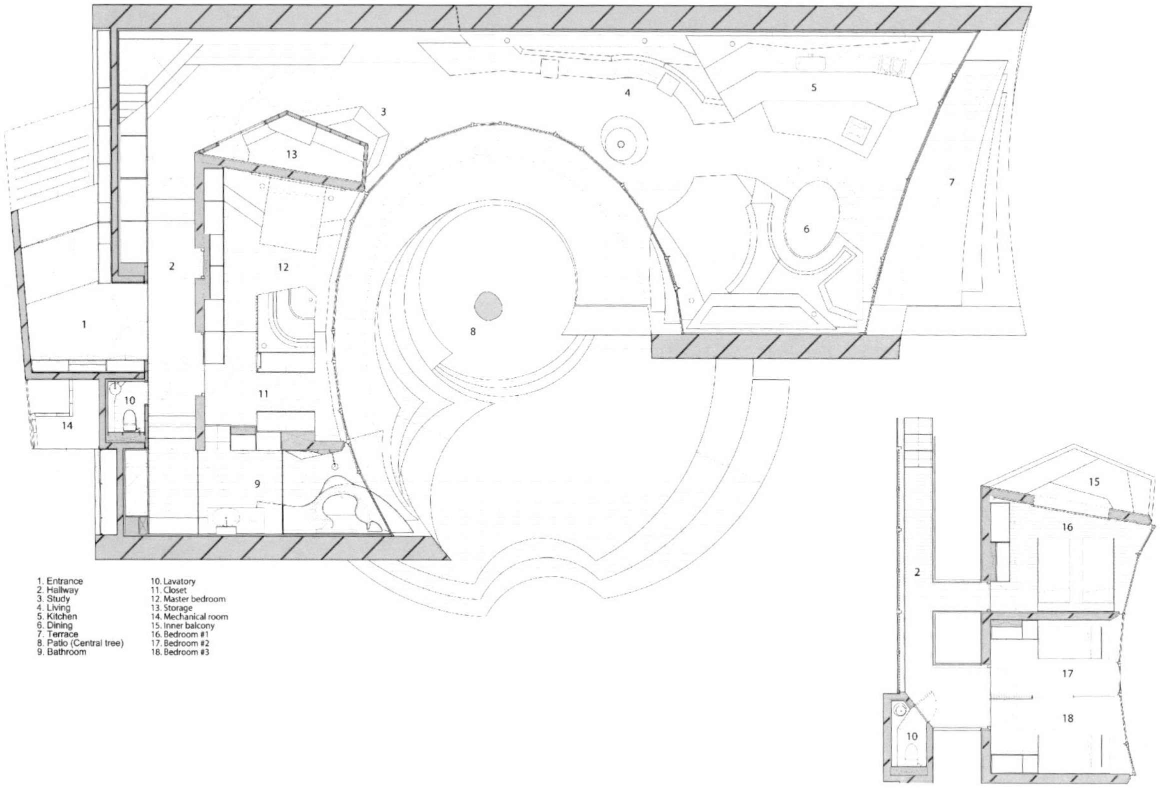

Plan/平面图

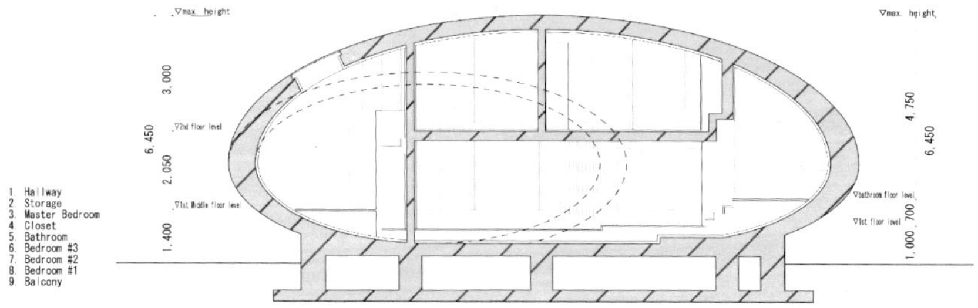

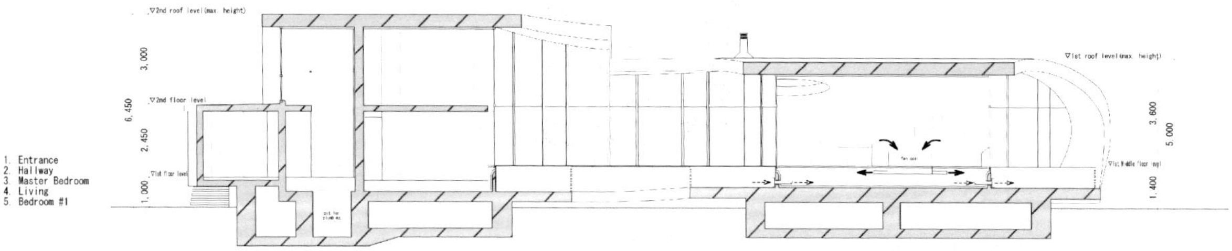

Section/剖面图

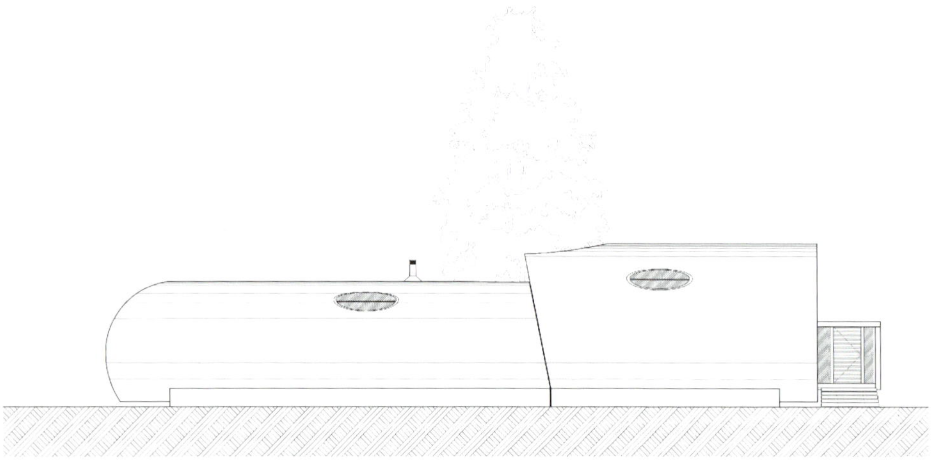

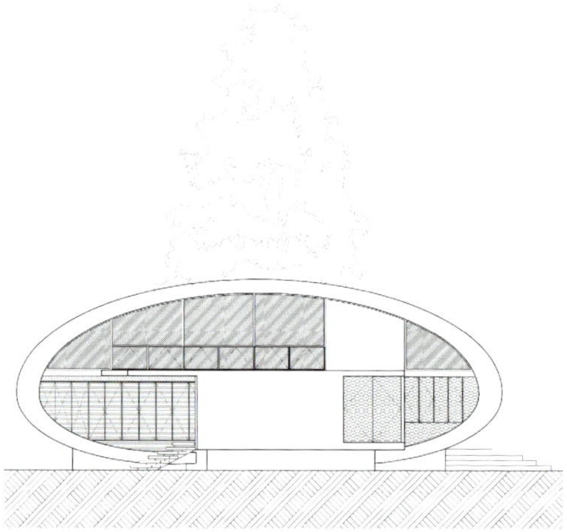

Facade/立面图

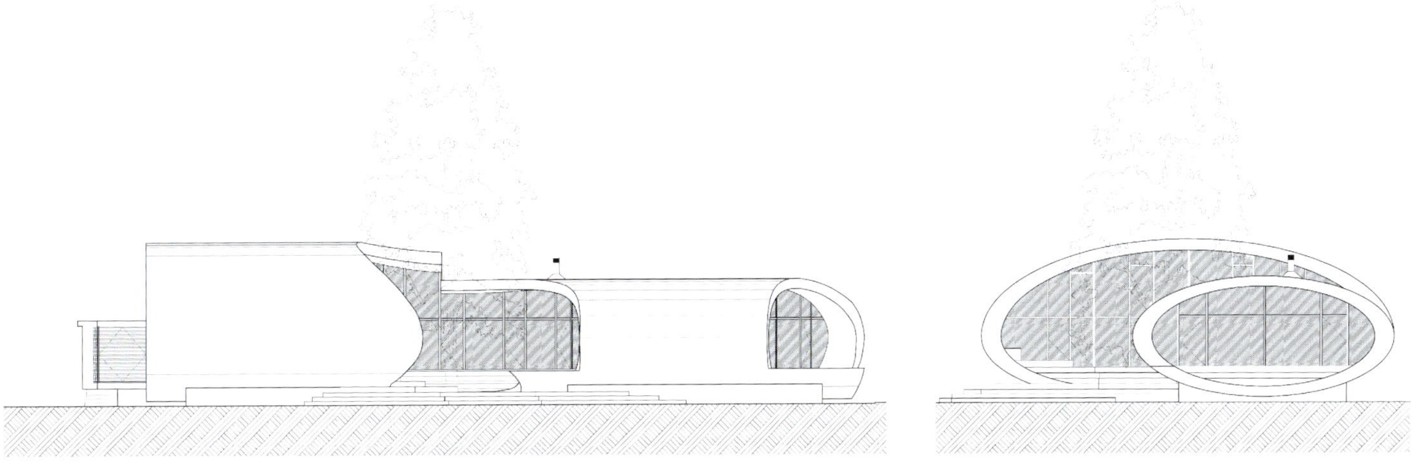

Facade/立面图

M&M Rosie

ARTechnic architects

- Architecture Design/建筑设计: ARTechnic architects
- Project Architect/项目建筑师: Kotaro Ide, Ruri Mitsuyasu
- Location/地点: Tokyo, Japan
- Area/面积: site 232 m² / total floor 151 m²
- Completion/建成时间: 2010
- Photograph/摄影: Nacasa & Partners Inc.

It was a strip of pentagonal land facing west with only a pair of parallel boundary lines. The land was such an odd shape, we were sure that we could create a house with more freedom from the extra space by giving a careful attention to improve the land's disadvantages such as an angle, orientation, height limitation and confined facade.

Two parallel boundary lines became the main axis, the angle of the front street, west boundary line and angles perpendicular to it became the other axes. We created the facade with a gate like structure with eaves of a roof slab and side walls to face straight to the street. Due to the consideration of sunlight for the neighbors who lived north of the house, we decided to bring the first floor lower than the land height in order to have enough ceiling height for the second floor. To handle this height matter (the building being lower than surrounding land), we have set the concrete intermediate floor slab to avoid the possibility of water coming into the interior space. The west side back yard was all decked, created to bring up the viewer's focal point level gradually from the interior floor, intermediate concrete floor to exterior deck for providing the exaggerated perspective view among the crowded neighborhood.

The reinforced concrete walls contained insulating concrete forms. The exterior walls were exposed concrete and finished with the penetrative sealer. In order to provide comfortable living, we placed the cold/warm air in-floor heating underneath the floor. The air circulated under the floor and finally came out through the grills located around the perimeter zones.

这块地坐东朝西，是一个有着一对平行边界线的五边形。面对这样一个奇特的地形，设计师认为，只要努力改善地形上的缺陷，如角度、朝向、高度的限制和狭窄的立面，就可以将房屋打造出更多的空间，从而提升居住的舒适度。

住宅的正面就像是一个门，用屋顶檐板和侧壁构造而成，直接面对街道。考虑到北面相邻住宅的采光问题，设计师决定让一层低于地平面，这样二层就有了足够的层高。为了处理这个高度的问题，设计师设置了混凝土的中间层厚板，这样水就不会流进室内。 西边的后院被设计为一个露天平台，能够使观赏者的视线从室内逐渐转移到混凝土的中间层，再转移到外面的甲板上，这样一来，在相对拥挤的周围环境中，设计仍能创造出一种夸张的透视效果。

钢筋混凝土墙能有效地隔热，外墙上覆盖着混凝土、胶粘剂和密封剂。为了打造舒适的环境，设计师采用了地板采暖，冷空气或暖空气在地板下面循环流动起来，最终透过格栅扩散出来。

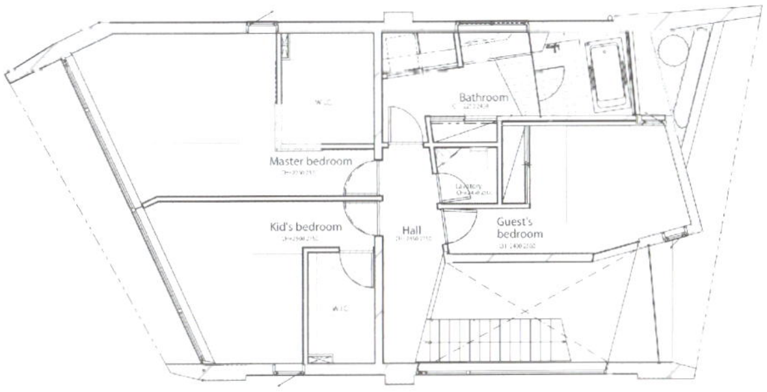

Second Floor Plan/二层平面图

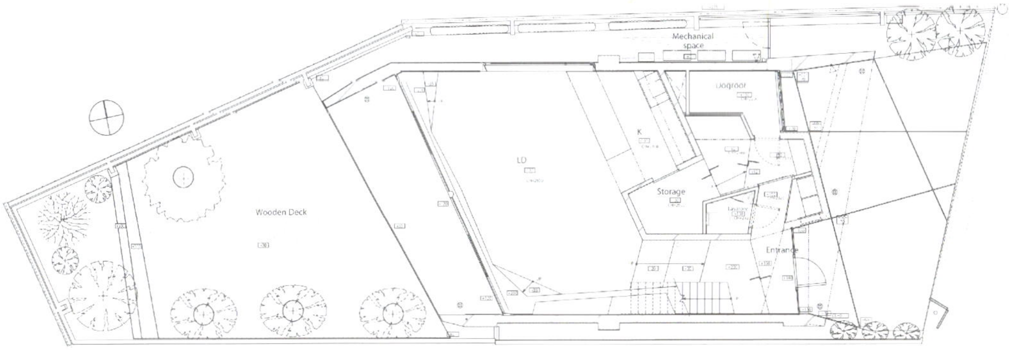

First Floor Plan/一层平面图

Lien Residence

Ministry of Design

■ Architecture Design/建筑设计: Ministry of Design ■ Project Architect/项目设计师: Colin Seah, Park ■ Location/地点: Singapore ■ Area/面积: 600 m² ■ Completion/建成时间: 2009
■ Photograph/摄影: Edward Hendricks , Patrick Bingham-Hall

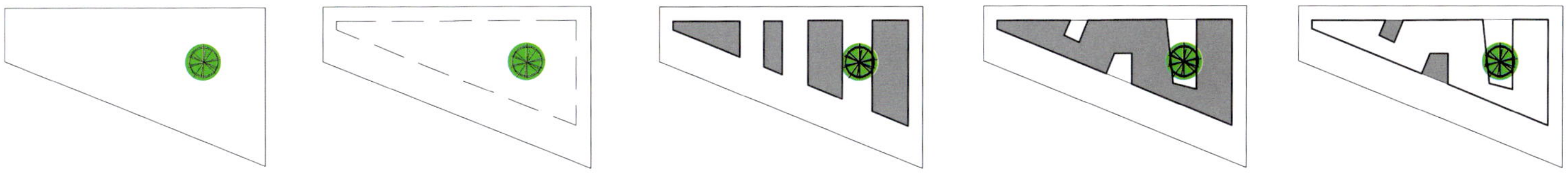

Returning to the romance of the single storey bungalow house, this zig-zag house acquires its unique form via a series of formal maneuvers around a mature tree located on its long and triangulated sliver of land.

Tropically acclimated to the region, the building's twisting form creates "in-between" spaces which provide shelter from nature's harsh elements and simultaneously allow for cross ventilation and filtered light. Courtyards, captured by the turning of the twisted building form, bring light into the basement service areas. Internal corridors serve as breezeways between air-conditioned and naturally cooled areas.

Slightly lofted over the ground, each of the building's three Miesian inspired wings house an entertainment zone, a family zone and a private master zone. Seen as a seamless singular form, the building reads as both sheltering building as well as abstracted sculpture. Viewed from the vicinity's taller structures, the building's roofscape provides the final design touch – where diagonally arranged planting strips echo the unique twisted form of the House Around a Tree.

通过采用一系列井井有条的措施，设计师将这栋房屋建在一个狭长的三角地带，环绕在一棵枝繁叶茂的大树周围，这样的条件构成了它现在独有的扭曲造型，令人仿佛回归到独栋别墅的浪漫气息中。

鉴于该地区的热带气候特点，该建筑的扭曲造型构建了"介于两者之间"的空间，为对抗大自然的恶劣天气提供了避风港，同时能够交叉通风、过滤光线。虽然庭院被扭曲的建筑造型所占据，但好处是自然光线可以照到地下室区域。室内通道充当起了空调与天然荫凉区域之间的通风走廊。

三个密斯风格的配房，略高出地面，分别构成了一个娱乐区、起居区和隐秘的主人活动区。这个看似完美的独特造型建筑，即可当作遮风挡雨的房屋，又可作为一个抽象的雕塑。从附近高处俯视，该建筑的屋顶景色为该设计定下了最终的格调——在以对角线划分出的区域分别种植绿化带的方式，与这栋房屋的扭曲造型交相辉映。

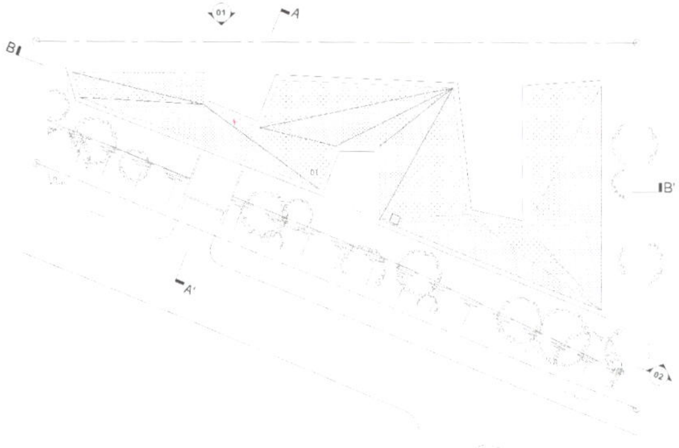

Roof Plan/屋顶平面图

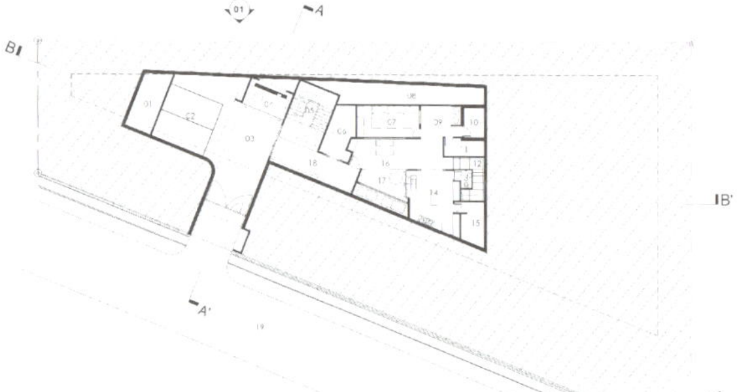

01 M&E SERVICES
02 PARKING SPACE
03 DRIVEWAY
04 ENTRANCE LANDSCAPE FEATURE
05 MAIN STAIRCASE
06 OUTDOOR EQUIPMENT ROOM
07 MAIDS ROOM
08 M&E SERVICES (POOL)
09 UTILITY AREA
10 HOUSEHOLD SHELTER
11 UTILITY STORE
12 SERVICE STAIRCASE
13 BASEMENT TOILET
14 WET KITCHEN
15 DRY STORE
16 INFORMAL DINING SPACE
17 SUNKEN LANDSCAPE COURTYARD
18 ENTRANCE LOBBY
19 HOLLAND PARK ROAD

Basement Plan/地下一层平面图

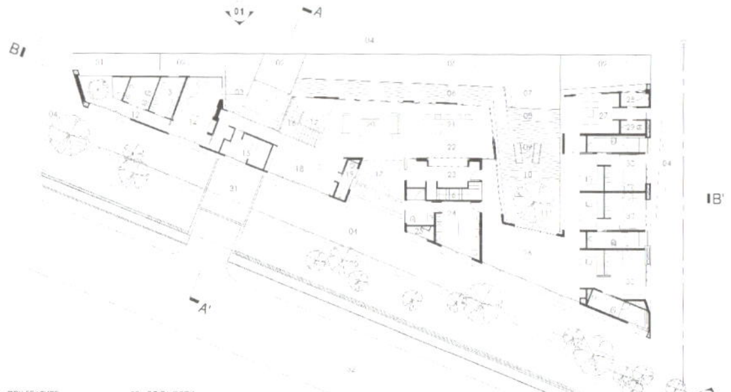

01 DRY FEATURE
02 WATER FEATURE
03 OUTDOOR DECK
04 GARDEN LEVEL
05 SWIMMING POOL
06 COVERED POOL DECK
07 POOL LOUNGING AREA
08 LOWER POOL DECK
09 POOL DECK
10 BBQ AREA
11 LANDSCAPED COURTYARD SPACE
12 MASTER BATHROOM
13 WALK-IN WARDROBE
14 MASTER BEDROOM
15 STUDY
16 STAIRS
17 VOID
18 BAR
19 POWDER ROOM
20 LIVING SPACE
21 DINING SPACE
22 BREAKFAST BAR
23 DRY KITCHEN
24 IN LAW'S SUITE
25 STORE
26 LIBRARY & MUSIC ROOM
27 INDOOR/ OUTDOOR FAMILY ROOM
28 SHOWER ROOM
29 POWDER ROOM 02
30 BEDROOM
31 GRAVEL GARDEN ABOVE ROOF OF DRIVEWAY
32 HOLLAND PARK ROAD

First Floor Plan/一层平面图

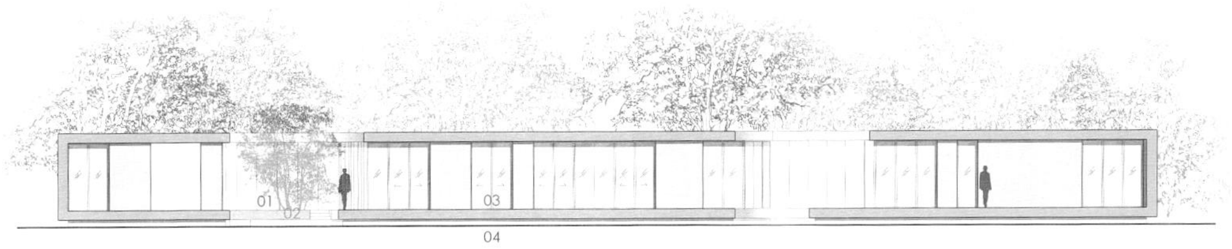

01 POOL DECK
02 LOWER POOL DECK
03 COVERED POOL DECK
04 SWIMMING POOL

Facade/立面图

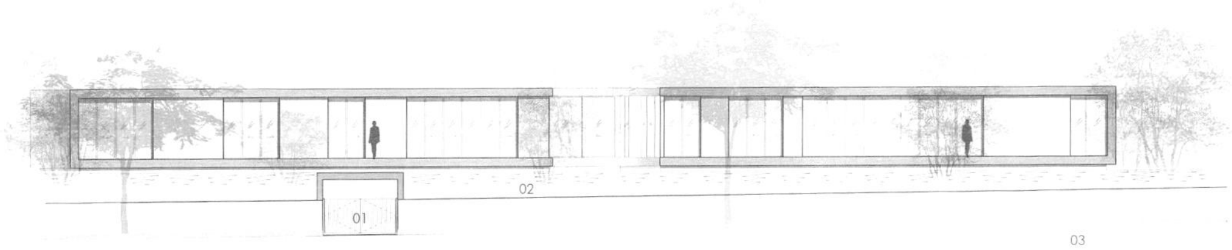

01 ENTRY CULVERT
02 GARDEN
03 HOLLAND PARK ROAD

Facade/立面图

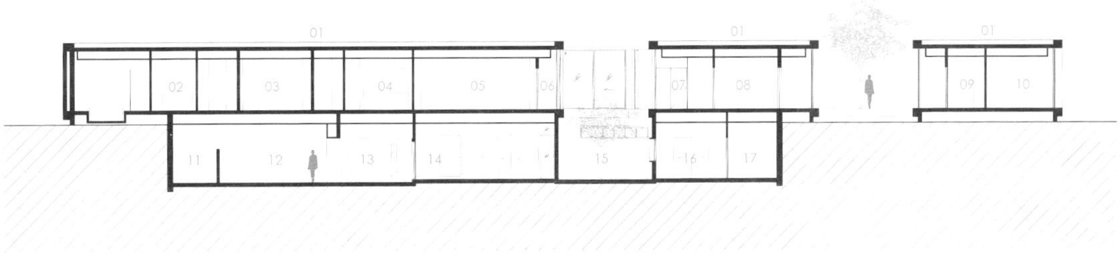

01 TURFED ROOF
02 MASTER BATHROOM
03 MASTER BEDROOM
04 STUDY
05 BAR AREA
06 POWDER ROOM
07 IN LAW'S SUITE BATHROOM
08 IN LAW'S SUITE
09 STUDY 02
10 BEDROOM 02
11 M&E SERVICES
12 PARKING SPACE
13 DRIVE WAY
14 ENTRANCE LOBBY
15 SUNKEN LANDSCAPED COURTYARD
16 WET KITCHEN
17 DRY STORE

Section/剖面图

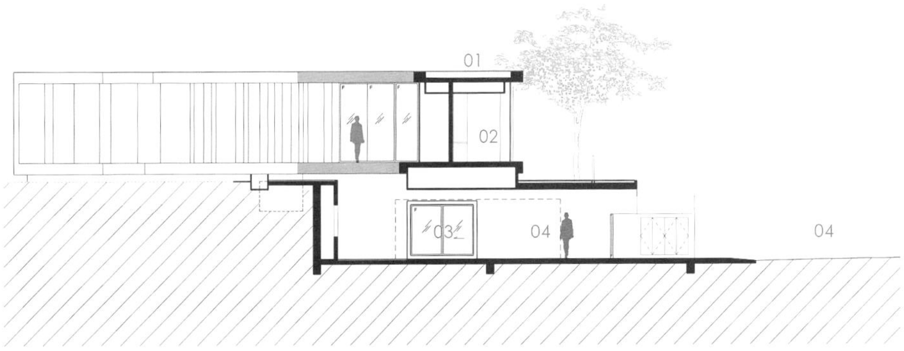

01 TURFED ROOF
02 STUDY
03 MAIN DOOR
04 DRIVE WAY/ CARPARK
05 HOLLARD PARK ROAD

Section/剖面图

Ontario Residence

Ministry of Design

■ Architecture Design/建筑设计: Ministry of Design ■ Project Architect/项目设计师: Park ■ Location/地点: Singapore ■ Area/面积: 420 m² ■ Completion/建成时间: 2009
■ Photograph/摄影: Edward Hendricks

Inspired by a series of challenging situations, the Ontario Residence by Ministry of Design seeks relevant and authentic solutions that challenge prevailing conventions of local luxury bungalow design.

The first challenge is rooted in the context of site - "How to create a sense of privacy to a house that has little visual privacy from its neighbours in spite of sitting on its own piece of land?" The second is the awkwardness of the car porch which typically manifests itself as a standalone or disjointed object – "How to incorporate it seamlessly into the language of the overall building?"

The primary part of the building is built on an assemblage of simple geometries: a vertical tower block juxtaposed with a horizontal block, capped by an overhanging hood – the car porch. The main public spaces are turned inwards towards a double-storey courtyard and lap pool. Organised around this central space, the activities of the residents enjoy a sense of clam privacy bathed in captured natural light from above. The tower houses the master wing, which is constantly cooled by the prevailing cross-breezes.

An oasis for its residents, the Ontario Residence is characterized by graceful proportions, pure geometries and a sense of reserved materiality.

项目Ontario Residence的灵感源于Ministry of Design公司试图利用切实可行的方式挑战当地奢华别墅的主流模式。

第一项挑战是基于该项目周边的环境，即如何将房子建得具有隐秘性，即使站在邻居的院子里也能保证这种隐秘性。第二项挑战是如何改变不雅观的车库。当地的车库大多暴露在院子里，但往往与院子的风格脱节。因此，要考虑如何将车库与院子的主题风格完美地融合在一起。

建筑物的主体部分建在一个简单的几何造型上：一个垂直的塔式建筑与一个水平建筑结合在一起，伸出的部分成为车库的顶棚。主要的公共区域则朝内转向一个双层的庭院和一个小型健身游泳池。在这片中央区域，主人都可以在阳光下进行各项活动，享受独处的宁静。这座塔式建筑的主体，南北通风效果佳，因此可以经常保持室内凉爽。

本项目以优雅的外形、简单的几何造型和建筑的实质意义为特点，成为主人生活中的一片绿洲。

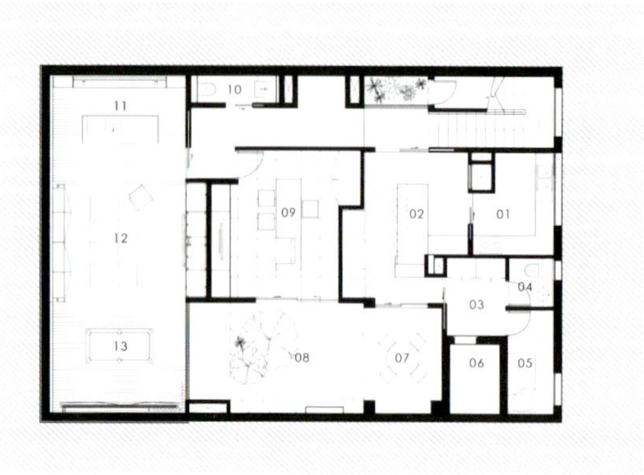

01 WET KITCHEN
02 DRY KITCHEN
03 UTILITY AREA
04 BATHROOM
05 MAID'S ROOM
06 HOUSEHOLD SHELTER
07 INFORMAL DINING
08 SUNKEN LANDSCAPED COURTYARD
09 STUDY
10 POWDER
11 BAR
12 ENTERTAINMENT ROOM
13 POOL TABLE

Basement Plan/地下一层平面图

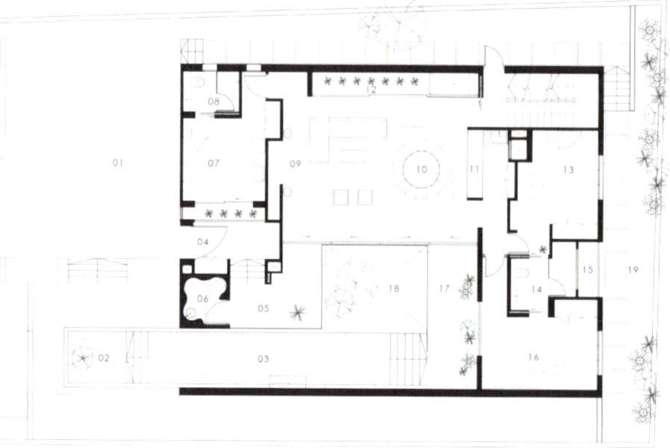

01 CAR PORCH
02 WATER FEATURE
03 POOL
04 ENTRANCE
05 ENTRY COURT
06 POWDER
07 BEDROOM 01
08 BATHROOM
09 LIVING
10 DINING
11 BREAKFAST COUNTER
12 PLANTER
13 BEDROOM 02
14 BATHROOM
15 PLANTER
16 BEDROOM 03
17 POOL DECK
18 VOID TO BASEMENT COURTYARD
19 BACKYARD / DRYING AREA

First Floor Plan/一层平面图

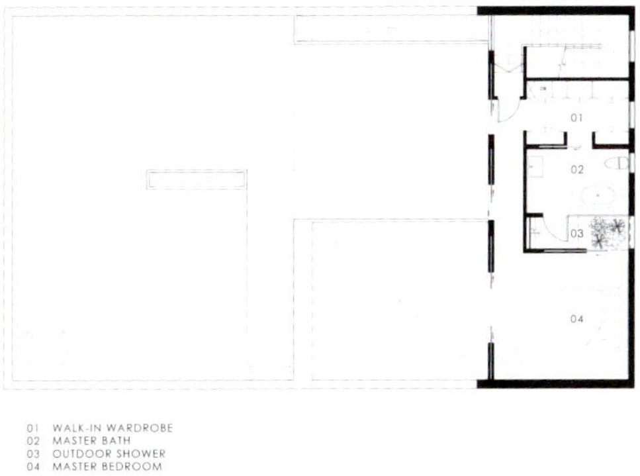

Second Floor Plan/二层平面图

Facade/立面图

27 East Sussex Lane

ONG&ONG Pte Ltd

■ Architecture Design/建筑设计: ONG&ONG Pte Ltd ■ Project Architect/项目建筑师: Maria Arango, Diego Molina ■ Location/地点: 27 East Sussex Lane, Singapore ■ Area/面积: 562.14 m^2
■ Completion/建成时间: 2008 ■ Photograph/摄影: Derek Swalwell

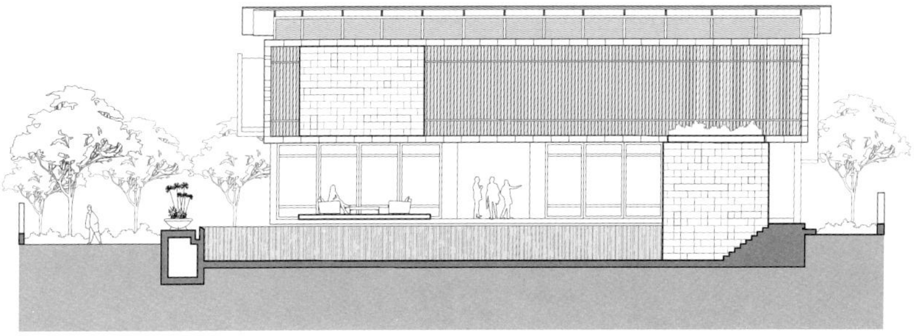

The external façade of 27 East Sussex Lane bears an uncanny resemblance to those of resort villas, with its grand entrance of stone and timber. Two stories make up the house, which also has a basement, and swimming pool in the back yard.

Natural textures feature prominently in the house, with dark, stained timber strips making up the surrounding fence, and long timber louvers framing the house's windows. Bamboo shrouds the house in mystery, giving off an oriental feel while also affording privacy.

Indoor and outdoor spaces blend together seamlessly, for example, the living room which leads straight to the pool via a stone veranda. From the second floor, the master bedroom stretches out into the backyard and over the pool.

With its use of diverse natural materials and the clever fusion of spaces, 27 East Sussex Lane is a delightful abode for rest and relaxation.

27 East Sussex Lane项目的外表看起来与那种度假别墅有着超乎寻常的相似,大厅入口的路都由石块和木条铺成。这个房子有两层，此外，还有一个地下室和后院的一个游泳池。

在这个设计中，自然纹路的应用非常突出，褪了色的木条拼凑成房子周围的栅栏，长条形的木条用作房子的天窗。在竹子的半遮半掩下，房子显得特别神秘,在保证了隐秘性的同时，散发着一种东方式的魅力。

室内与室外完美地融合在一起，例如，客厅通过石台，能够一直延伸到水池；而在二层，主卧伸展到了后院的水池上空。

设计师通过使用不同种类的天然材料，与自然空间巧妙地结合，使本项目成为一个休息放松的良好住所。

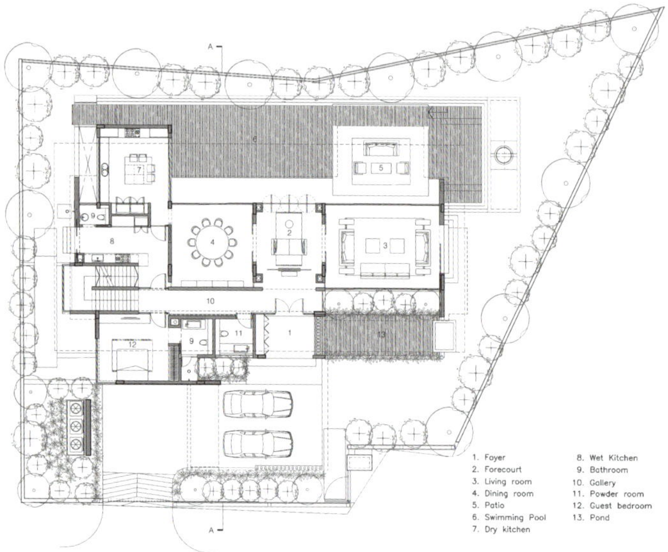

First Floor Plan/一层平面图

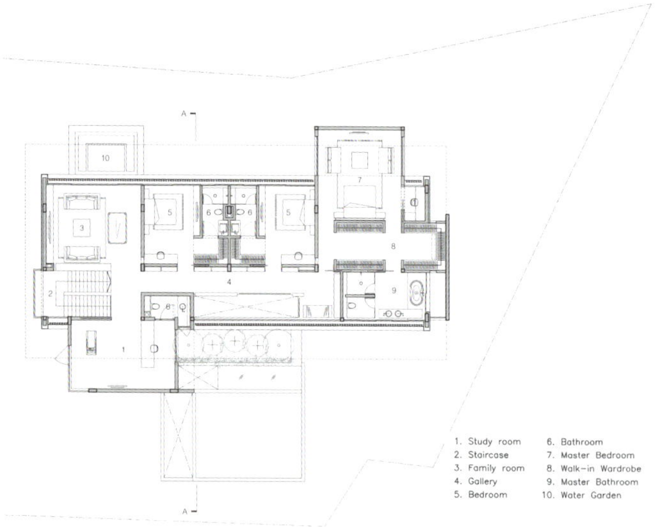

Second Floor Plan/二层平面图

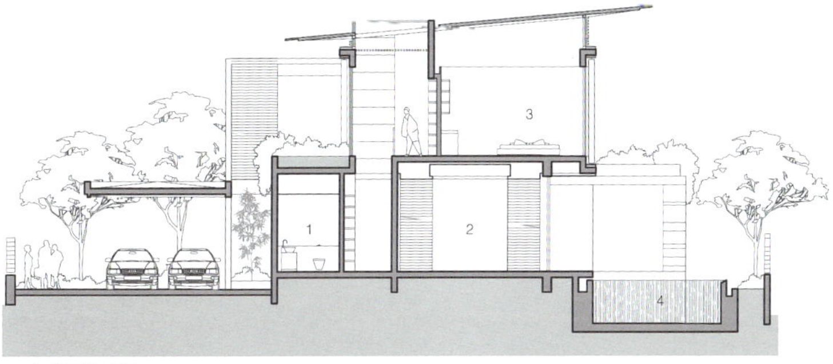

1. Powder room
2. Dining room
3. Bedroom
4. Swimming Pool

Section/剖面图

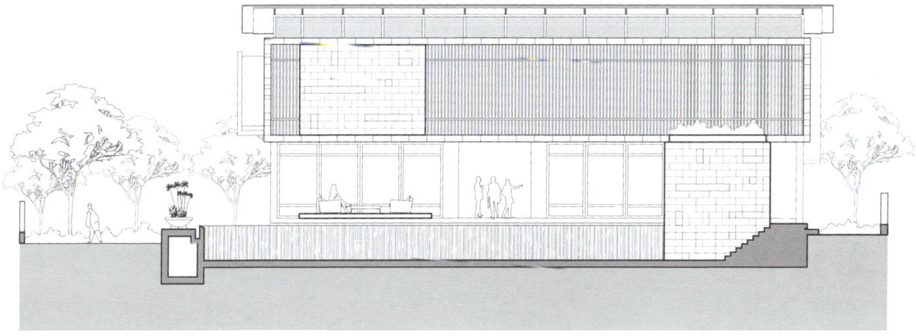

Facade/立面图

72 Sentosa Cove

ONG&ONG Pte Ltd

■ Architecture Design/建筑设计: ONG&ONG Pte Ltd ■ Project Architect/项目建筑师: Maria Arango, Diego Molina ■ Location/地点: 72 Sentosa Cove Drive, Singapore ■ Completion/建成时间: 2009
■ Photograph/摄影: Derek Swalwell

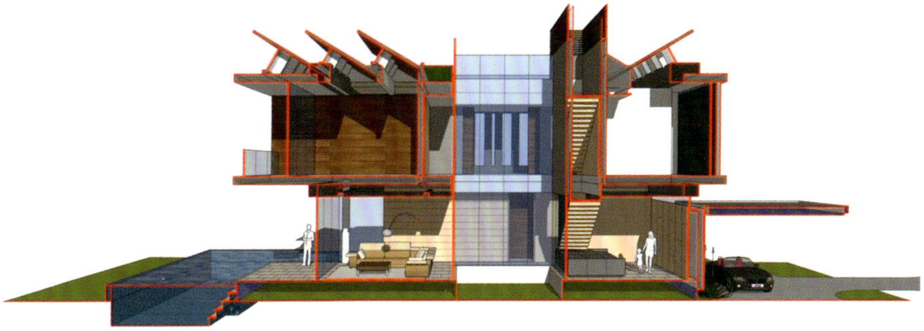

This is a home that can endure the tropical climate with minimal environmental impact, by tapping on available natural resources. Four levels make up the building, each catering to the family's various social and entertainment needs while also accommodating each member's individual need for solitude.

Achieving space maximisation and privacy were also factored into this design, in view of the house's close proximity to neighboring units. As a result, the internal spaces face inward into a multi-functional courtyard which acts as circulation pivot, light well and ventilation exhaust, while also serving as a visual focal point. Dense foliage also provides a natural means for additional privacy.

The building's slimmer east and west facades are a contrast to the wide-open north and south ones. This regulates natural lighting and wind ventilation for efficient thermal circulation. A pitched roof, with its series of repeating slopes, also generates additional skylight openings. And let's also not forget that this is first and foremost a home for an active couple and their kids. For this, a lap pool and training room ensure that their fitness needs are taken care of.

这栋房屋的设计师，仅仅通过利用当地自然资源的方式，就将热带气候对房子的影响降到最低。房子共有四层，每层都充分考虑家庭各种社交活动和娱乐的需要，同时满足了每位成员拥有独立空间的要求。

设计师在实现空间最大化的同时，也考虑到了靠近周边建筑的房间的隐秘性设计。最终，室内空间面朝多功能的庭院成为建筑的循环中枢、进光口、通风排气口，当然，室内空间也成为设计的一个亮点。

房子东西两侧单薄的墙面与南北两面宽阔的空间呈现出对比的效果。这样的设计可以调节自然采光，也可以帮助运行中的热循环系统通风换气。成排的倾斜屋顶能够提供更多的天窗。由于这里居住的是一对有小孩、有活力的夫妻，因此，这里还需要有一个小型的健身游泳池和训练室，从而保证全家的身体健康。

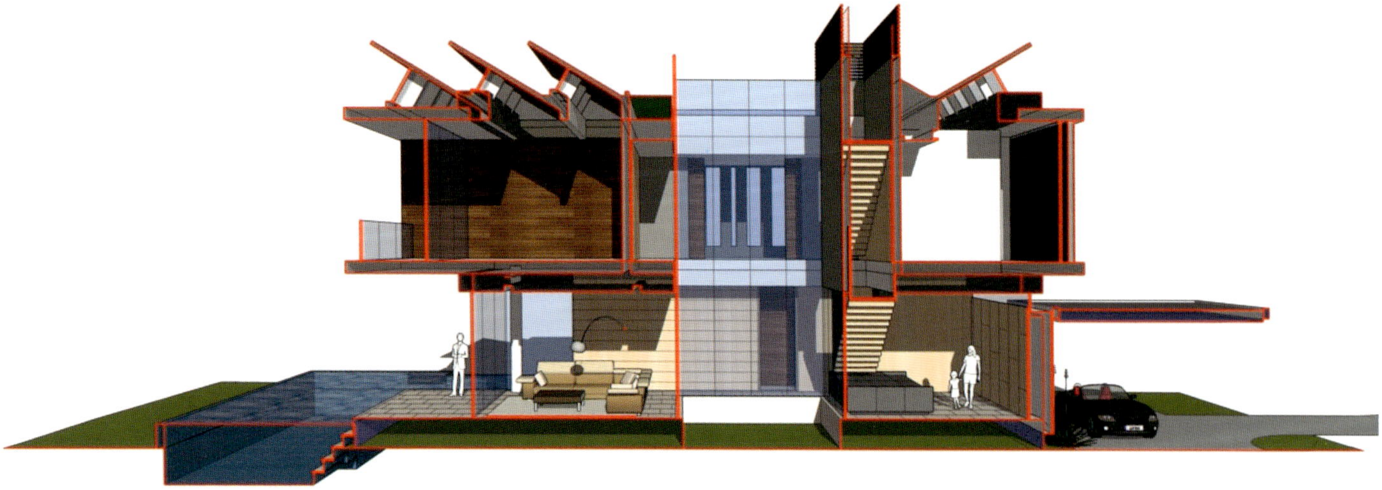

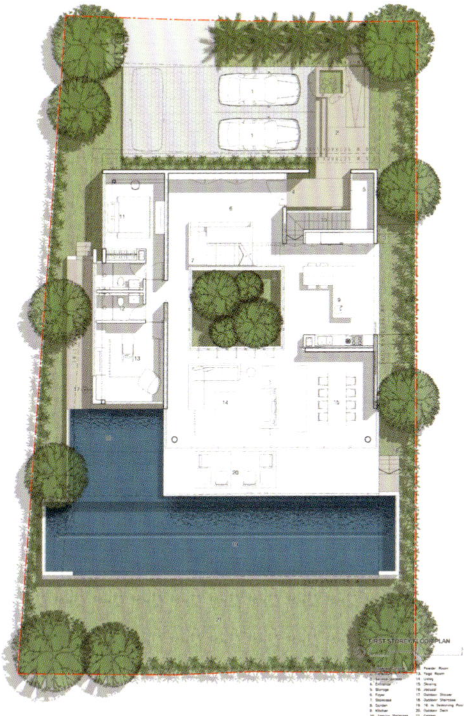

First Floor Plan/一层平面图

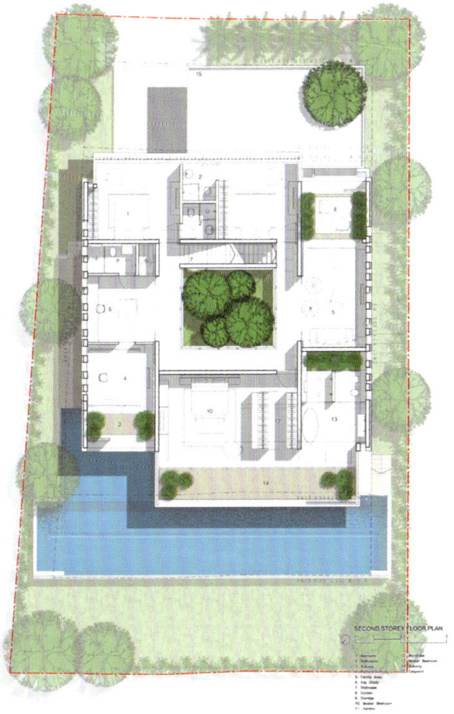

Second Floor Plan/二层平面图

CBS Building

George Dasic Architects

- Architecture Design/建筑设计: George Dasic Architects
- Project Architect/项目建筑师: George Dasic, Yushi Kobayashi , Toru Usui, Yuki Miyoshi Nikitaki, Tamaki Suzuki
- Location/地点: Tokyo, Japan
- Area/面积: site 93.19 m^2/building 49.09 m^2/floor 196.36 m^2
- Completion/建成时间: 2008
- Photograph/摄影: Peter Cook

As is often the case the seemingly contradictory requirements of regulation and requirements eventually led to thedesign concept.

We decided to pull the building perimeter further away from the street than regulation required creating a sense ofpavement and open space, more associated with European urban fabric and unique in this area. Since the seminarbusiness has frequent visitors arriving and leaving the building it allowed for a "transition" space. Furthermore, itavoided feeling of competing with cars for the entrance, so common within back streets of Tokyo.

Increased setback reduced the footprint of the building, but then the volume allowance permitted four comfortableheight floors. This matched the 4 main functions (Gallery, Office, Seminars and Living Pod) contained within 3 programmatic elements (Gallery, Business, Pod). This followed a concept of 3 floating cubes, stacked one above the other in the order of "programmatic hierarchy".

Cube 1 Containing the Gallery on Level 1 with a separate entrance; Cube 2 Containing the Business space (Office on Level 2 and Seminar room on Level 3); Cube 3 Containing the Living Pod, with a living area, kitchen and bathroom, on Level 4.

常常，表面上看似矛盾的常规要求，最终却会启发设计构思。

设计师决定将建筑物修建得比原计划离街道更远一些，这样就腾出了一小块类似于人行道的开放空间，使其更接近欧洲的城市结构风格，并在这里显得独一无二。由于这里开展研讨班的缘故，大批的访客会来来往往，于是这里便成为一个必要的过渡空间。这样的设计可以避免行人和机动车争抢通道，要知道这种情况在东京的小街道是屡见不鲜的。

自上而下，建筑物外围的缩进程度逐渐增加，这样有效地减少了该建筑的占地面积，但这样的空间只适合设计四层的建筑结构，这恰巧与四个主要功能房（画廊、办公室、研究室和活动室）相匹配，其中涵盖了三个计划设计的元素（艺术、工作和生活）。设计师依据打造三个漂浮的立方体的理念，将其一个接一个地按照“设计要求等级”堆砌起来。

第一个空间是一个有单独入口的一层画廊，第二个空间是商务区（二层是办公室，三层是研究室），第三个空间是生活区（在四层有一个客厅、厨房和卫生间）。

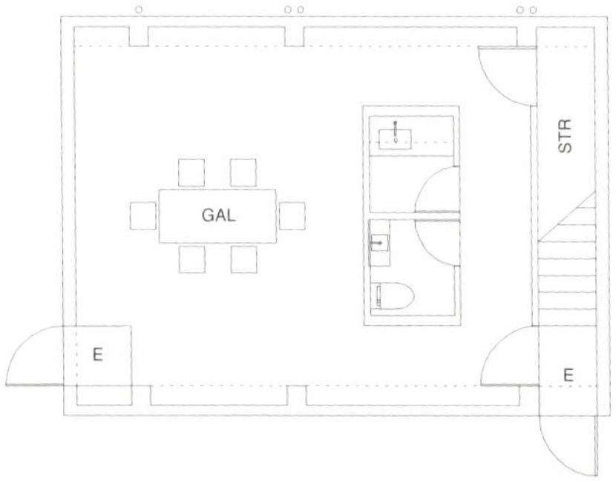

First Floor Plan/一层平面图

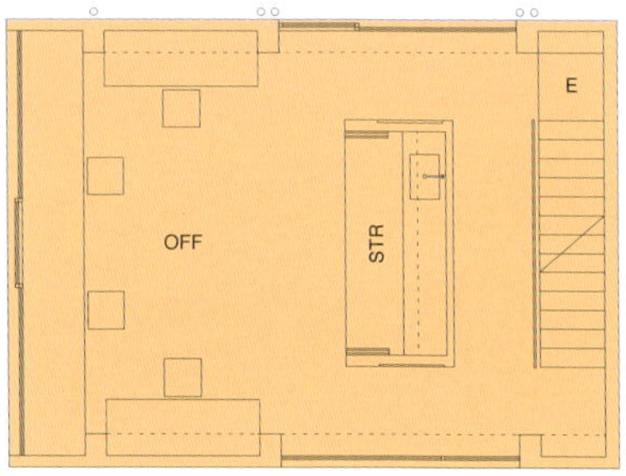

Second Floor Plan/二层平面图

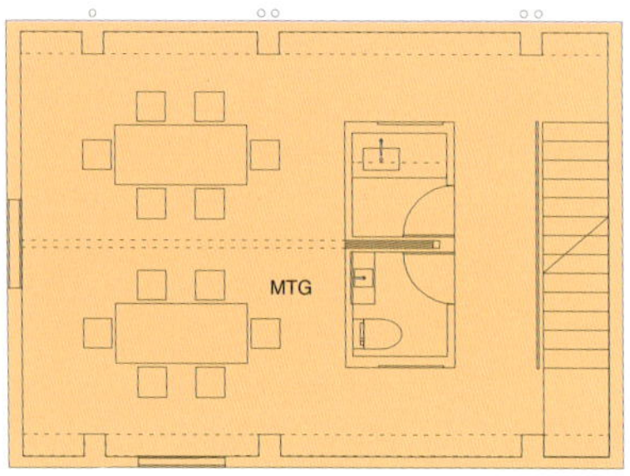

Third Floor Plan/三层平面图

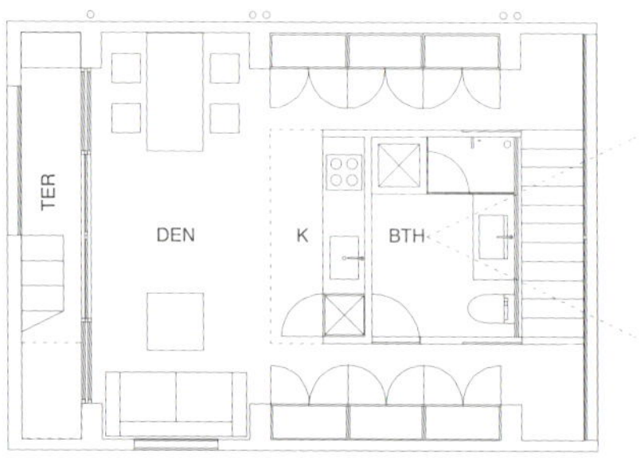

Fourth Floor Plan/四层平面图

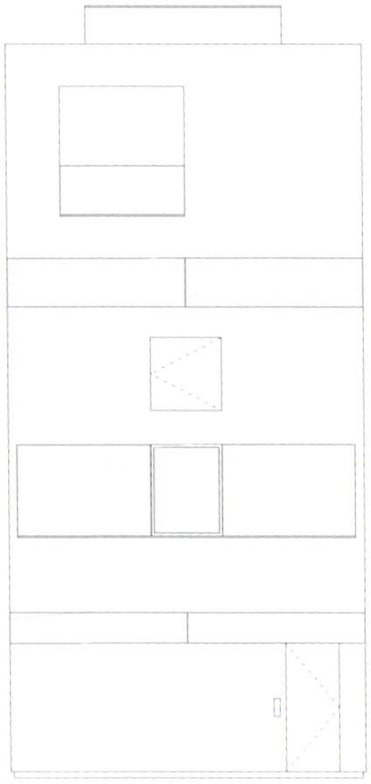
Facade: West/西立面图

Facade: South/南立面图

Facade: East/东立面图

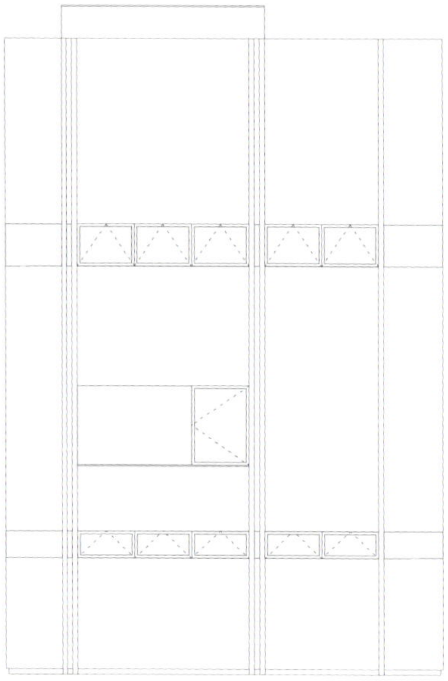
Facade: North/北立面图

3645 Minami Hara

George Dasic Architects

■ Architecture Design/建筑设计: George Dasic Architects ■ Project Architect/项目建筑师: George Dasic, Yushi Kobayashi, Yuki Miyoshi Nikitaki, Toru Usui, Tamaki Suzuki ■ Location/地点: Nagano, Japan
■ Area/面积: site 1,953.22 m^2/building 240.30 m^2/floor 196.74 m^2 ■ Completion/建成时间: 2008 ■ Photograph/摄影: Hiroyasu Sakaguchi, Arrow Resort Corporation

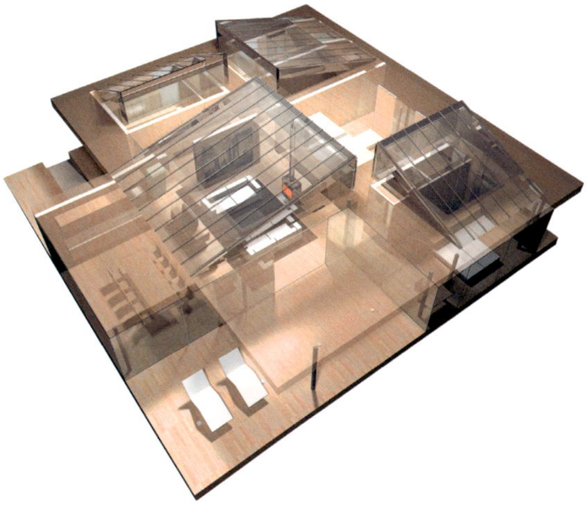

The site is located in the wooded area of Karuizawa resort town 150km away from Tokyo. It is part of thedevelopment consisting of 5 villas, each designed by different architect. The mountainous area has cool climate insummer and cold sharp winters. There is lots of rain in the summer but winters tend to be dry without much snowfall. Humidity and condensation are important factors.

Our design concept is based on the following principles: (1)Response to the Site and the Surrounding Nature. House is build on the top of the hill with each room taking advantage of the best views and orientation. (2)Sculptural, Geometrical Forms. Intention was to design a "sculptural object" which will stands as a counter balance against the backdrop ofnatural green forest and blue sky. (3)Space Flexibility and Fluidity / Parametric Layout. The overall space is designed as a fully fluid environment allowing for a sense of openness, lightness and easymovement. (4) Transparency. It wouldmake the building extremely transparent "merging" it with the surrounding forest, while the special site positioningwould preserve sense of privacy from the other four buildings within the development. (5)Building Methods and Tools . The main material used to clad both interior and exterior of the building is cheep and readily available Shinagohan panels that are usually used as an insulation material. (6)Sustainability and Environment – natural sunlight intake. Karuizawa planning law, amended in 2007 called for every new building to follow set of "aesthetic" rules includingsloped roof with eaves. we decided to use it as a vehicle to create naturally cooled, aired and heated environment with full intake of natural light. (7)Sustainability and Environment -air exchange. Second role of the shape and positions of the roofs is to allow for the natural air exchange and cooling in thesummer. (8)Barrier-free Access to every room and space. From the very beginning we decided to designed house, which will have only one level. (9)Dancing rain windows. The plan and the roof configuration positions one of the main windows on each side as a direct "slide" of the roofsloped plane.

这个建筑位于丰茂的轻井泽旅游胜地，距离东京150千米。该项目是住宅小区的一部分，而该住宅小区由5栋别墅组成，每一栋都分别由不同的建筑师设计完成。夏日里，该山区清爽怡人，雨水充沛；而到冬季，该山区则寒冷刺骨，干燥少雪。因此，湿度是影响该项目的重要因素。

设计师的设计理念主要基于以下几个原则：（1）与所处的地理位置和自然环境相呼应。房子建在山顶，每间卧室都能充分利用绝好的视野和方向。（2）运用雕塑和几何造型。这种想法是希望将建筑设计成为一种雕塑物体，能够使其与绿树、蓝天相平衡。（3）展示空间的灵活性和流动性（参数化布局）。整体空间被设计成完全流动的环境，营造出一种开放、轻巧、易于活动的感觉。（4）注重透明度。这将使建筑很容易融入周围的森林中，而该建筑所处的独特位置，依然可以与小区中的其他四栋建筑物相对保持隐秘性。（5）保证建筑方法和工具的使用。建筑物的内部与外部均使用常作为隔热材料的复合材料面板，这种面板廉价并且很容易买到。（6）坚持可持续性、环保性。轻井泽地区2007年修订的规划法案，要求每一栋新建的建筑物都要遵循“美学”规则，包括带屋檐的斜屋顶。由此，设计师决定利用这一方式，创造出天然的制冷、通风和供暖系统，并创造出可以全面采光的环境。（7）坚持可持续性、环保性理念下的自然换气、通风。屋顶的位置和形状起到的第二个作用，就是在夏日里可实现自然通风和制冷。（8）房间和空间的通道中的无障碍穿行。从一开始我们就决定将这个房子设计为只有一层的建筑物。（9）设置观赏雨窗。这个计划和屋顶的结构，决定了建筑每一扇都要安装一面大窗户，这样一来，雨水就会从倾斜的屋顶滑落至此，使看着雨景的主人乐在其中。

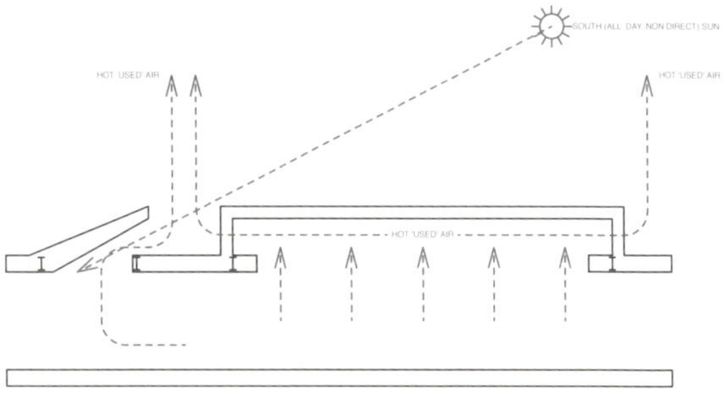
SOUTH (ALL DAY NON DIRECT) SUN
HOT 'USED' AIR
HOT 'USED' AIR
HOT 'USED' AIR

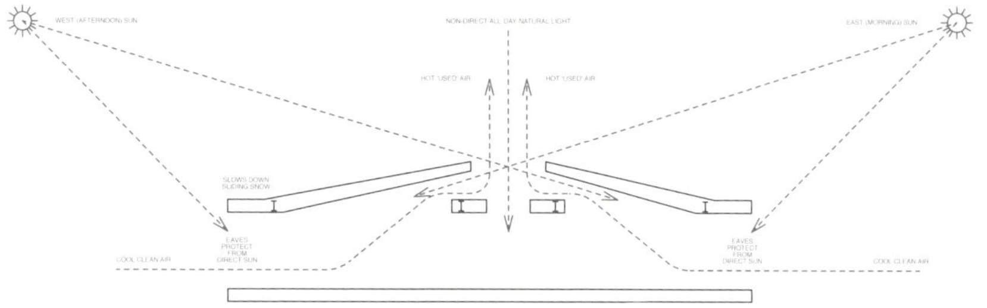
WEST (AFTERNOON) SUN
NON-DIRECT ALL DAY NATURAL LIGHT
EAST (MORNING) SUN
HOT 'USED' AIR
HOT 'USED' AIR
SLOWS DOWN SLIDING SNOW
EAVES PROTECT FROM DIRECT SUN
EAVES PROTECT FROM DIRECT SUN
COOL CLEAN AIR
COOL CLEAN AIR

Section/剖面图

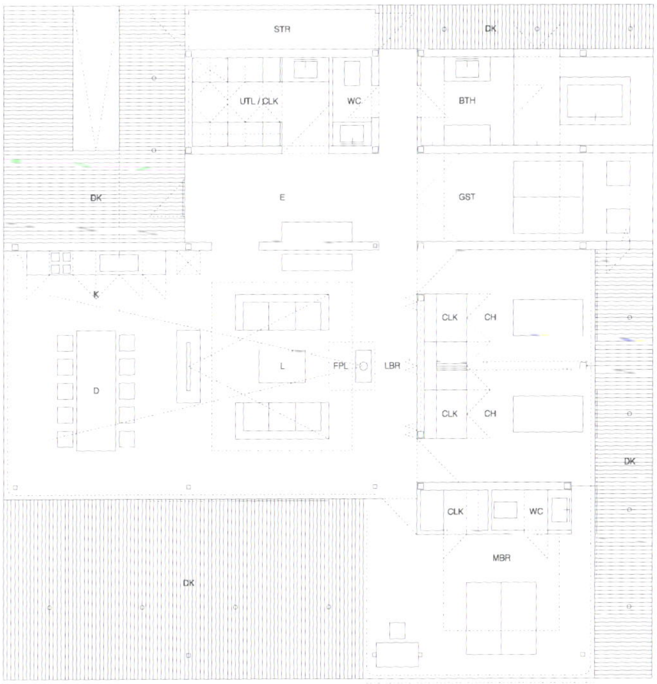
STR
DK
UTL / CLK
WC
BTH
DK
E
GST
K
CLK
CH
D
L
FPL
LBR
CLK
CH
DK
CLK
WC
MBR
DK

Plan/平面图

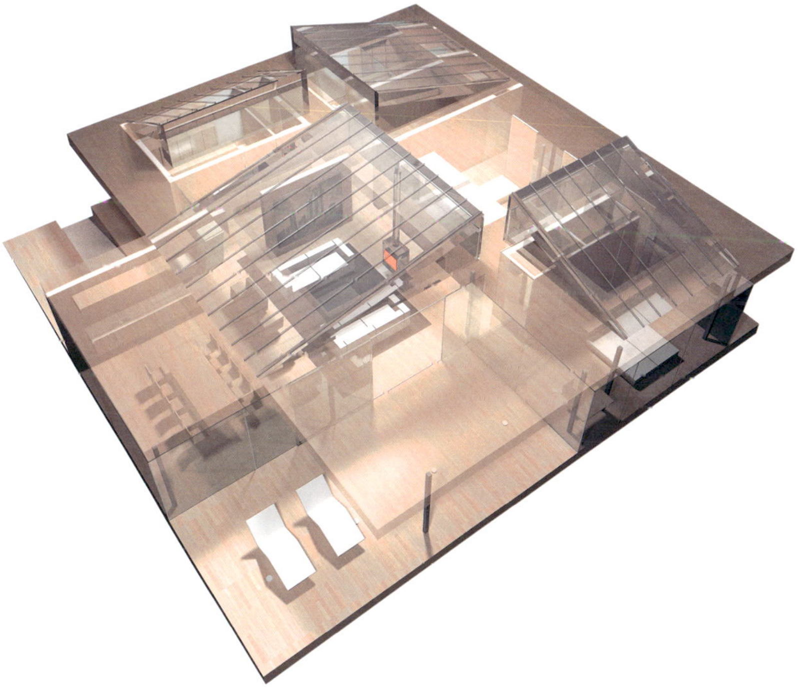

Yoyogi House

frontoffice/tokyo

- Architecture Design/建筑设计: frontoffice/tokyo
- Project Architect /项目建筑师: Koen Klinkers, Will Galloway, Iwao Takeuchi
- Location/地点: Tokyo, Japan
- Area/面积: site 89.68 m^2/ building 78 m^2/gross 99 m^2
- Completion/建成时间: 2008
- Photograph/摄影: Daisuke Akita

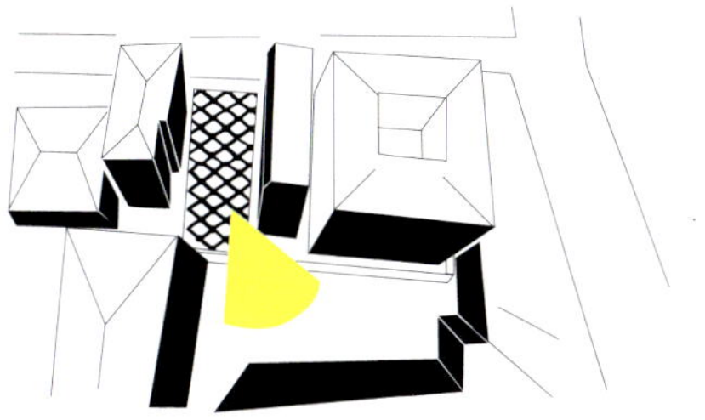

a long narrow site, without external views and a limited wedge of solar access

the site becomes a kind of isolated valley

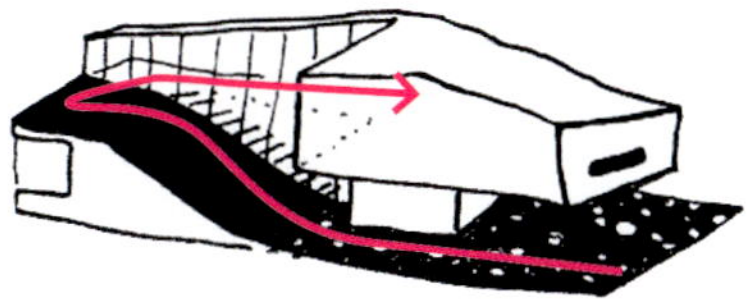

making landscape as essential as interior space

The site we were presented with by the client is typical of Tokyo's older residential areas - defined by narrow streets whose character emerges more from the individual choices of the homeowners than from any comprehensive urban plan. More specific, the elongated and narrow plot (5.4m x 16m) is surrounded on all sides by houses built to the razor's edge of their property lines and to the maximum height the local zoning laws allow. The result: a site flanked by tall blank walls on its two long sides, and facing directly onto a neighbour's home to the South, with only a sliver of open sky available on the south-east corner.

The clients, a middle-aged couple, wanted a small and efficient home that would accommodate their desire for personal space - for each to pursue their own hobbies, and especially a small painting area and work station for the wife. They also asked for access to outdoor space that could be used easily in daily life, and a garden where strawberries could be planted to indulge in a longtime hobby of making home-made jam.

Surprisingly the home feels larger than its 80m^2 would suggest, due in no small part to the long views outside and the active green space at the center of the home. The house is treated in much the same way, fitting into the context without attempting to alter it, mimic it, or deny it.

客户提供给我们的这个建筑位于东京典型的老居民区内，街道狭小，多是房主们自己造成的，而不是城市规划的。更具体一些，就是说在长而窄小的区域(5.4米×16米)布满了大大小小的建筑，达到了当地法律允许的极限。结果导致这栋房子两侧高大的白墙直接面对南向的邻居家，在房子里只能看到东南角的一线天空。

客户是一对中年夫妇，希望能有一栋面积虽小但使用率高的房子，两人都能有私人空间来做他们各自喜爱的事情，尤其是妻子希望能有一个作为她的工作间的小画室。他们还要求可以方便地进出于屋内和屋外，而室内空间必须能满足他们日常生活的需要，满足夫妻二人长久以来的喜好——亲手制作草莓酱。

令人意外的是这栋房子比之前所说的80平方米要显得大一些，因为从远处看来，房子没有琐碎的格局，而且房子的中央空间是明快的绿色，同时，这栋房屋的设计风格也充分融入了当地特色，而不是对其加以改变，模仿或是否定。

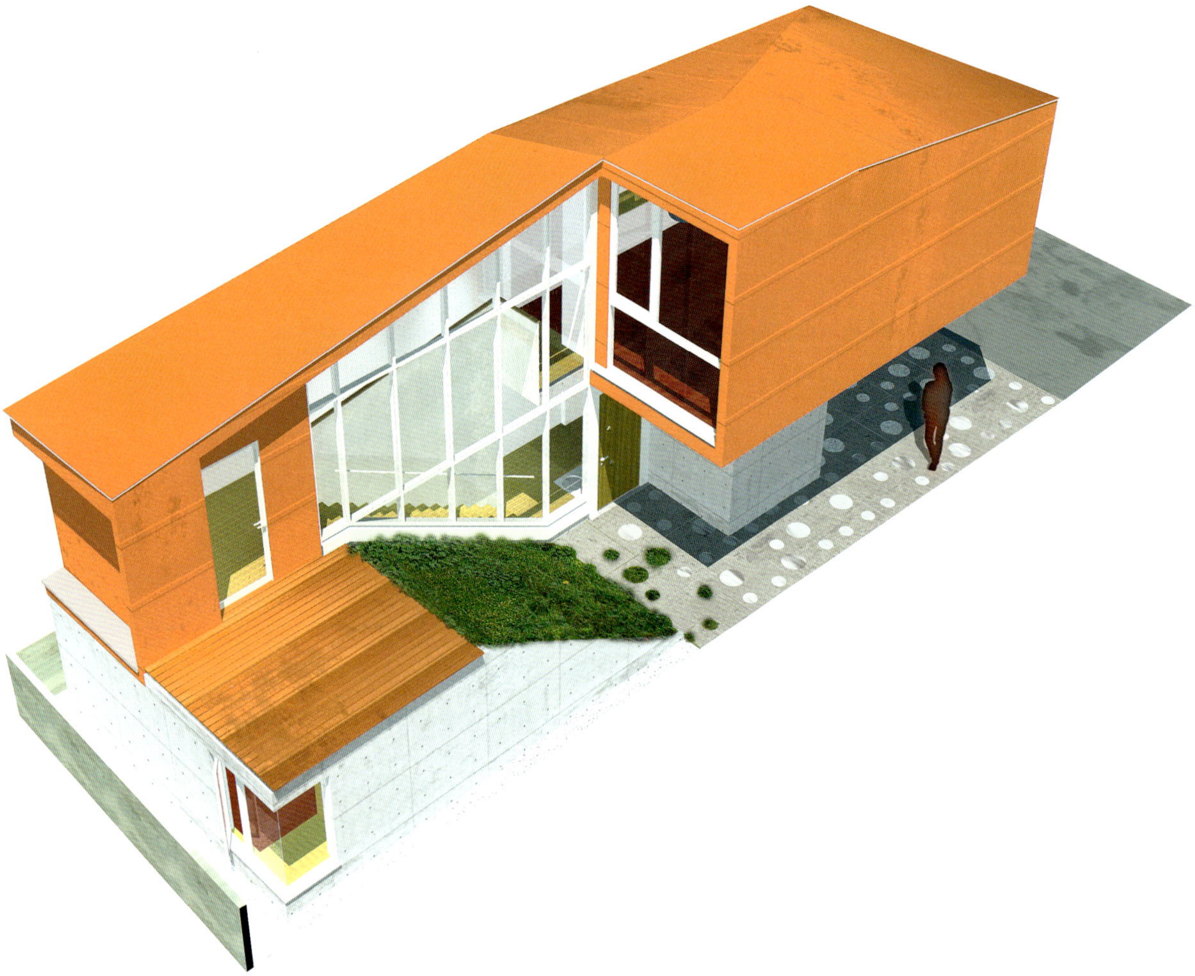

1 entry
2 bedroom
3 bath/laundry
4 wc
5 kitchen
6 living/dining
7 work space
8 parking
9 garden
10 deck

Plan/平面图

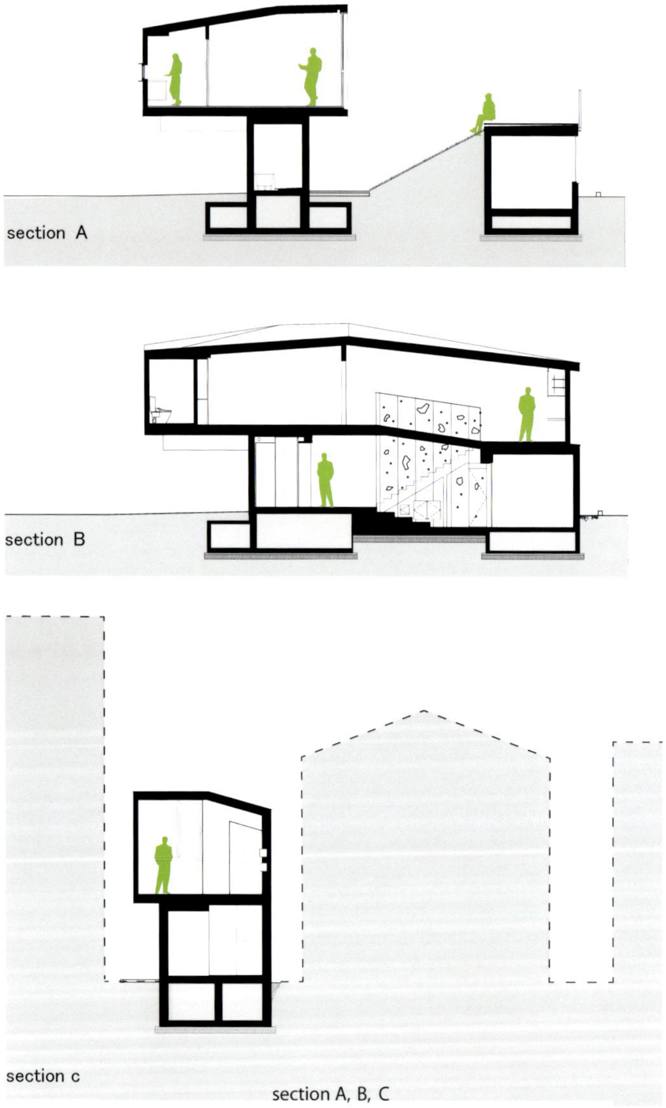

Section/剖面图

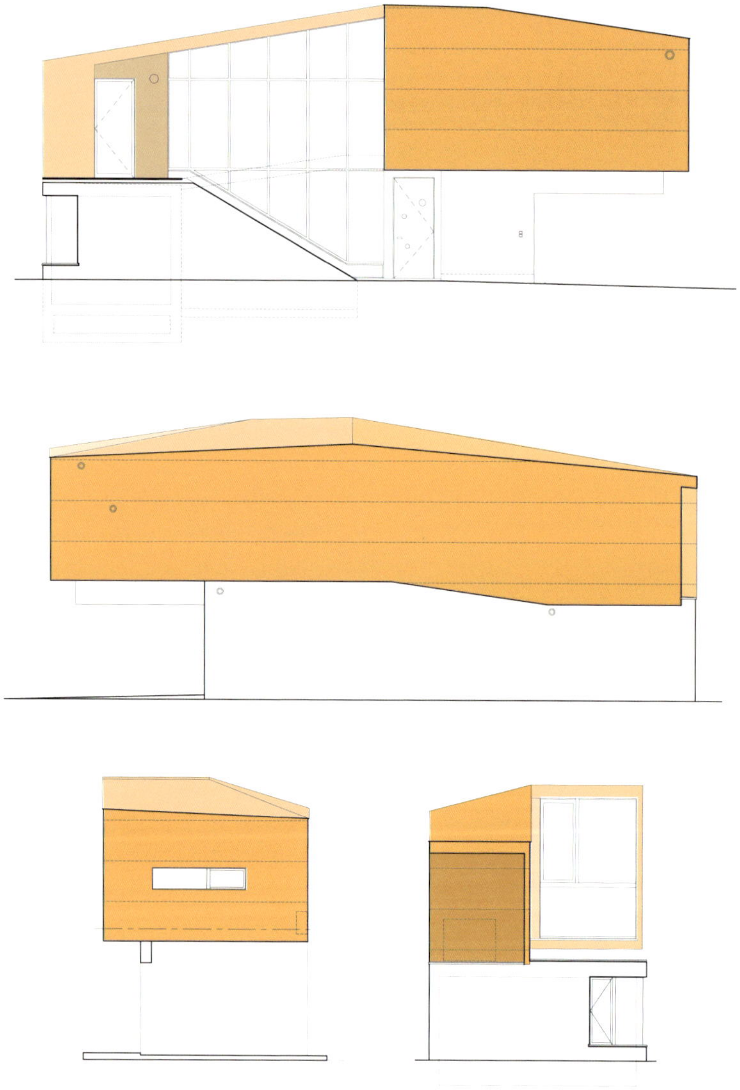

Facade/立面图

Minami-azabu House

frontoffice/tokyo

- Architecture Design/建筑设计: frontoffice/tokyo
- Project Architect/项目建筑师: Koen Klinkers, Will Galloway, Max Kim
- Location/地点: Tokyo, Japan
- Area/面积: site 142.94 m²/building 185.06 m²/ gross 270.78 m²
- Completion/建成时间: 2010
- Photograph/摄影: Daisuke Akita

In Japan the building code enforces an envelope under which any building must fit. The rules are designed to guarantee access to light for neighbours, which makes sense in a dense urban environment. Ironically, on a landlocked parcel of the type we were given the rules also ensure the lower levels of any building constructed will be submerged in shadow. It is in fact tempting to simply live in the dark. The building code also ensures the largest floor areas are available on the lowest floors, and the impulse is often to design a home so the living area will be as large as possible. Simply accepting the darkness is not uncommon, especially on small properties. The ambitious architect may work at mitigating the effect by building a deep light well to be certain residents will at least see the reflected light of the sun if not the sky during the day – a Tokyo lifestyle is as a result often dark and introverted.

Here we took a different tack, and reversed the standard typology – the main bedrooms are on the second floor, parking, a hard-surface landscape, and a guest room are on the ground, and the living areas are lifted to the top of the home where they are connected to the city by a series of decks and a stairway to a large roof deck. Since the ground level does not receive enough light to make a proper garden we chose to move the ground to the third floor and to the roof, planting it with gardens that will allow the owners of the home to have large green spaces in the centre of the city.

In a city like Tokyo space is a luxury. By keeping the ground level open the city becomes a kind of backdrop to the entrance. In the same way the decks and balconies that open the home to the city create an enormous outdoor space that is very nearly unheard of in such a site.

在日本，任何建筑必须遵守建筑法规，这些法规旨在保证邻居的采光度，这对于一个人口稠密的大城市来说，情有可原。然而，有讽刺意味的是，对于这样一个被包围起来的区域，该法规却允许任何建筑物的较低层都被淹没在阴影里。实际上，完全生活在黑暗中还是很有挑战的。建筑法规也确保最低的楼层可以获得最大的面积，因此设计师往往将房屋的起居面积设计得尽可能大，完全处于阴暗的情况时有发生，特别是那些小的住宅。有经验的建筑师会通过建造一个深而且明亮的楼梯井来减轻这些影响，至少这样可以保证那些白天看不到天空的的居民能看到反射的阳光，要知道东京的生活方式经常是黑暗和封闭的。

在这里，设计师采取了不同寻常的策略，即将主卧室都设在二楼，而停车场和客房都设置在一楼，起居区被安置在顶层。通过连续平台和楼梯，主人可直达屋顶的大平台，在那里，他们可以看到整个城市。由于底层采光不足，不能建造一个花园，因此设计师便将其移到三楼及屋顶，这样一来，房子的主人可以在城市的中心拥有大片的绿色空间了。

在东京这样的城市，空间是极其宝贵的。通过保持地面空间的开放，使城市成为该建筑入口处的一道背景。同样，平台和阳台的位置使整个住宅面向城市，为建筑创造出了巨大的室外空间，这几乎是很少见的。

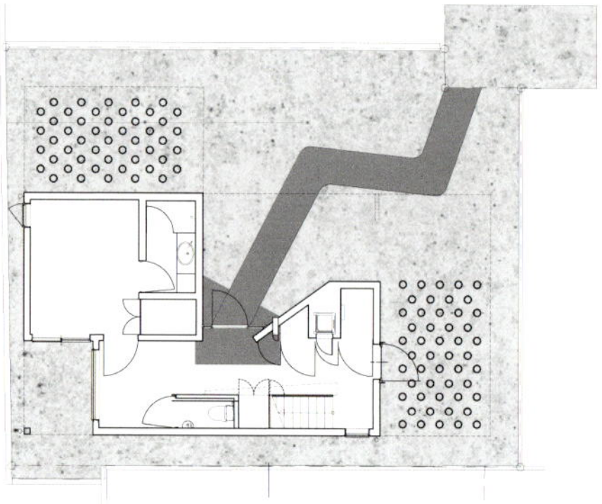

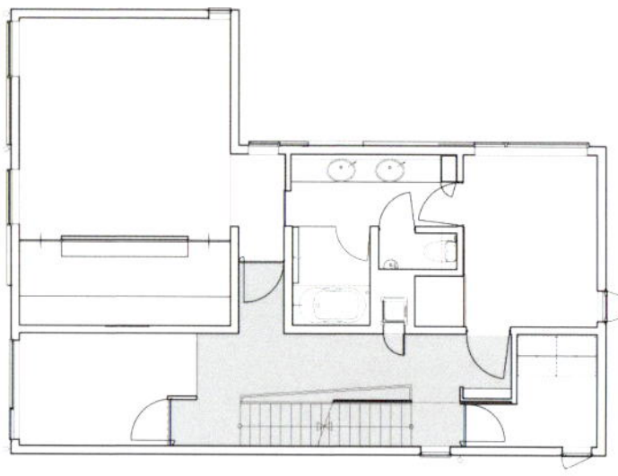

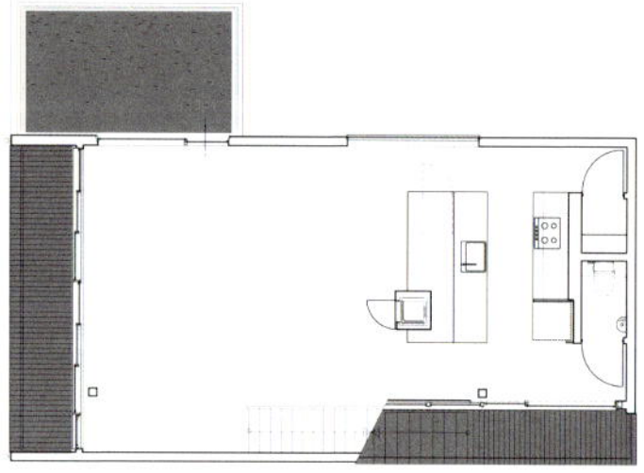

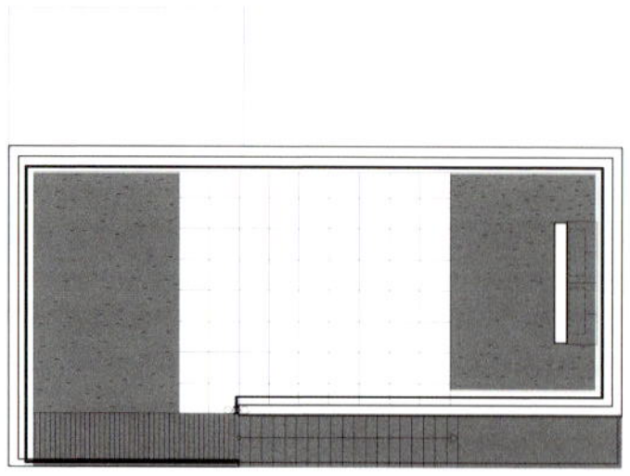

Plan/平面图

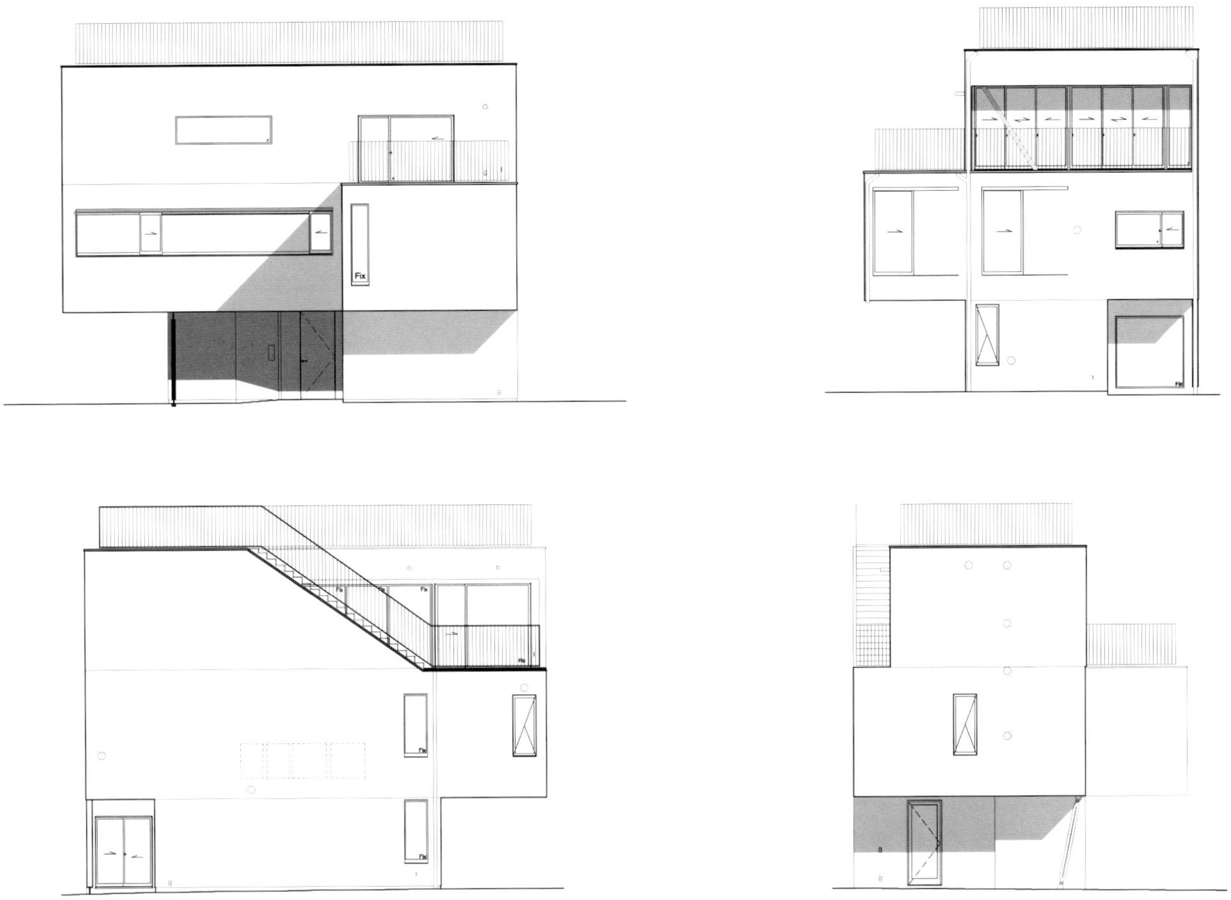

Facade/立面图

Section/剖面图

Acrylic House

TAKESHI HOSAKA ARCHITECTS

- Architecture Design/建筑设计: TAKESHI HOSAKA ARCHITECTS
- Project Architect/项目建筑师: Takeshi Hosaka
- Location/地点: Yamanashi, Japan
- Area/面积: site 1,024.47 m^2/building 73.28 m^2/ total floor 103.09 m^2
- Completion/建成时间: 2006
- Photograph/摄影: Toshihiro Sobajima

This building is a residence for a young husband and wife and their three children. It stands amid rice fields and houses, located a little off the main road for sightseeing. There are zelkova trees, silk trees, and vegetative soil on the property, and the surrounding nature, such as the view of the nearby mountains and the clean air, is very appealing. On the other hand, however, the attention from and the presence of the sightseeing road and scattered houses are out of harmony with the beautiful nature and therefore a little problematic. So, we have decided to design the house in such a way that the inside is always in contact with the outside, creating a sense of open space, yet it is hard to see what is going on inside the house from outside the property, making the boundary between the inside and the outside as "transparent" as possible. To be specific, we have made the extremely transparent, continuous boundaries between the inside and outside that don't have joints or mullions by using heat-insulating, 20mm-thick acrylic boards. Acrylic fiber has very few impurities, so the surface doesn't reflect much light even during the day, making it easy to see the inside from the outside quite transparently. Just as you can see the garden from the living room, kids playing in the garden can see the inside, too.

Each side has an opening for entering and exiting, making it easy to access the garden and the terrace from anywhere inside the house. The inside and the garden are divided only by the extremely transparent, continuous boundaries with no joints, it feels like they are more intermingled than before. It is a never-seen-before connection between the inside and the outside. I get excited just thinking what the inside of this building will be like amid the colorful summer blossoms, in the sunshine and the moonlight, in the rain, in the typhoon, amid the autumn leaves, and in the winter snow. There also exists a new relationship between human beings and nature here. We are hoping that we can achieve unprecedented transparency and a new connection between the inside and the outside that are not possible with glass constructions.

这栋房子是为一对年轻的夫妻和他们的三个孩子准备的。它建在稻田和其他房屋之间，离观光主路有些远。这里有光叶榉树、木棉树，也有一块可种植植物的土地，还有四周的自然风光，比如不远处的山峰，这些是多么引人入胜。另一方面，不管怎样，观光路上的风景、稀稀拉拉的屋舍与这迷人的自然环境是不和谐的。设计师决定将房子设计成为即使人在室内也可以与外界联系的形式，创造出开放的空间。然而，从房子外面很难了解室内的情况，设计师尽可能地将室内与室外的界限“透明化”。

为了突出这一特点，设计师将整栋房子设计得非常透明，利用 20毫米厚的丙烯酸隔热设施作为房间连续不断的分界线，使室内与室外找不到断点。丙烯酸纤维的杂质非常少，因此其表面很难反光，即使是白天，也非常容易从外面看清楚室内的情况。主人可以从客厅一直望到花园，孩子们也能在花园看到室内的一切。

建筑物的每一面都有出入口，这样可以更容易地穿过花园，进入房间。室内与花园仅仅是由没有断点的分界线分开，透明且连续不断，和过去相比，它们现在更好地融合在一起，这是前所未有的设计。设计师在思考：怎样设计建筑物的内部才会使人置身其中像簇拥在夏日的花丛里，沐浴在由日光和月光下，感受秋日的落叶、冬日飞扬的白雪。在人类与自然之间还有这样一种新的关系存在，设计师设想，可以不用玻璃结构连接室内室外，便可以创造出前所未有的透明感和连通感，他们的确做到了。

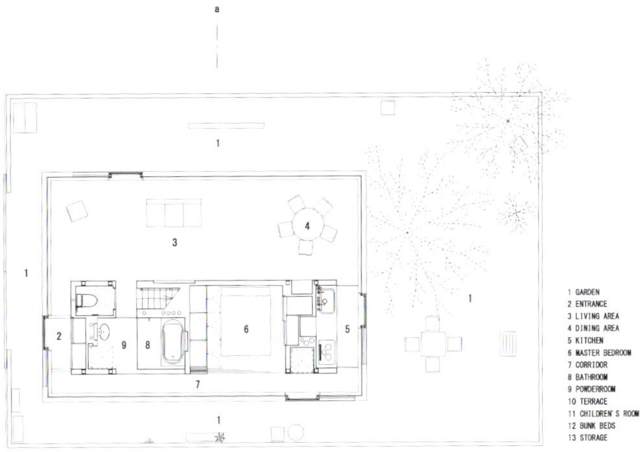

First Floor Plan/一层平面图

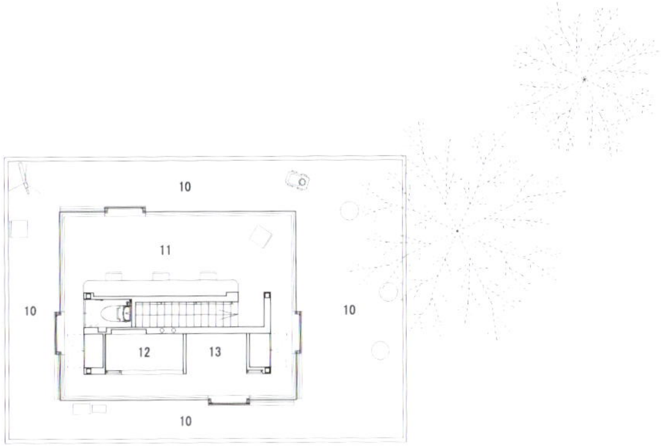

Second Floor Plan/二层平面图

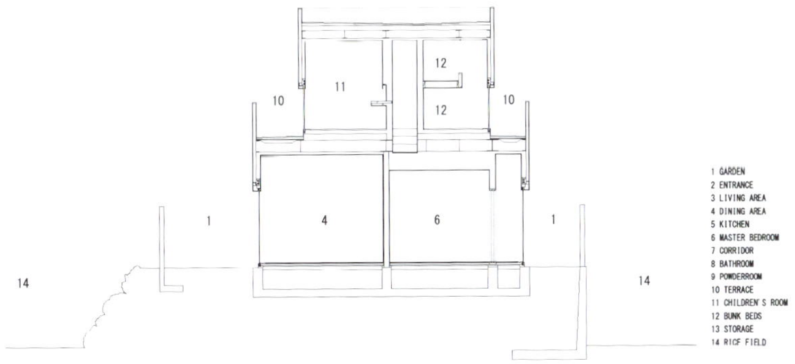

Section/剖面图

inside house & outside house

TAKESHI HOSAKA ARCHITECTS

- Architecture Design/建筑设计: TAKESHI HOSAKA ARCHITECTS
- Project Architect/项目建筑师: Takeshi Hosaka
- Location/地点: Tokyo, Japan
- Area/面积: site 95.98 m^2/building 37.32 m^2/floor 71.20 m^2
- Completion/建成时间: 2009
- Photograph/摄影: Masao Nishikawa

This is a residence consisting of inside house and outside house.
It is also a proposal of architecture of a new relationship between internal and external.
Because the site is an irregular shape, the planar shapes of the rooms in the two cubes of "inside house" and "outside house" are irregular shapes. In the two cubes, there are glass windows and wooden-board windows in the peripheries, and any room lets in light and wind from the windows and the inside house intermingles with outside house through windows. When the wooden board windows are opened, liberating level increases and light and wind further intermingle with surrounding environment. When the wooden board door at the entrance is opened, it is planned so that the view can run through from the north garden to the south garden and other wooden board windows are also planned so that the view can run through unexpectedly.
The "inside house" houses rooms where a family of four people can live in comfort, places equipped with water supply and electric appliances, etc. It consists of entrance, living room, dining room, TV, air conditioner, kitchen, bedrooms, children's room and bathroom, etc. The "outside house" houses lives that you want to do in the outside house than the inside house. For example, keywords such as woodwork, work, growing plants, keeping insects, maintenance of bicycles, hammock, sunbathing, gardening, playing with water, seeing the sky, tools for mountain climbing, napping and reading books, etc. are assumed. The opening sections can be opened with different heights and sizes between the inside house and outside house and they were planned so that you can feel more distant than the actual distance and various places can be generated in the small site.

有种住宅，涵盖了室内房和露天房两种。
同时它也提出了一种全新的室内室外的建筑关系。
由于这块地皮形状不规则，所以设计构建了两种形式。其中，外形平坦的房间是两个所谓的“室内房”，不规则的则是“露天房”。这两个室内房，外围有玻璃窗和木窗，透过窗，阳光和风能够进入每间卧室。正因为有了这些窗户，室内房与露天房才能很好地融合在一起。当打开木窗时，房间更加通透，阳光和风也融入其中。当打开入口处的木门时，能够欣赏到花园的全景，而从其他窗口也能看到超乎想象的风景。
“室内房”的房间配有供水、供电等设施，可以给一家四口提供非常舒适的居住环境。室内房有走廊、客厅、餐厅、厨房、卧室、孩子们的房间、卫生间以及电视、空调等。而室外房，则是为主人从事室外的活动而准备的。如木工活、劳动、种植、养昆虫、修自行车、吊床、日光浴、园艺、戏水、望远、户外装备、读书等等。室内房与露天房之间开放的部分可按需要调节，这样设计的目的是为了在这片空间内，给人一种更开阔、更丰富的感觉。

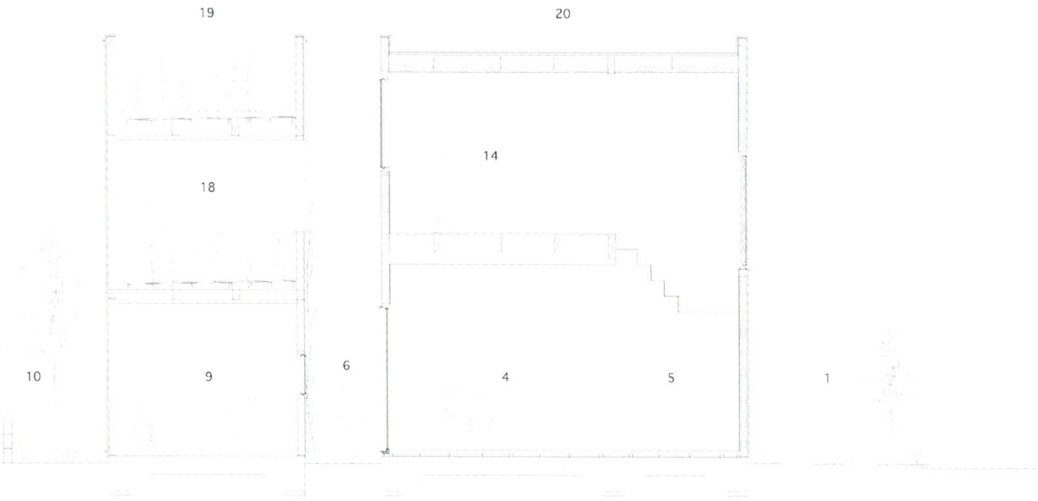

1:north garden
2:entrance
3:living
4:dining
5:kitchen
6:alley garden
7:Bicycle maintenance room
8:outer room
9:Atelier room
10:south garden
11:room
12:powder room
13:bath
14. bed room
15:clothet
16:hanmoc terrace
17:void
18:flower room
19:roof terrace
20:roof

Section/剖面图

1:north garden
2:entrance
3:living
4:dining
5:kitchen
6:alley garden
7:Bicycle maintenance room
8:outer room
9:Atelier room
10:south garden
11:room
12:powder room
13:bath
14: bed room
15:clothet
16:hanmoc terrace
17:void
18:flower room
19:roof terrace
20:roof

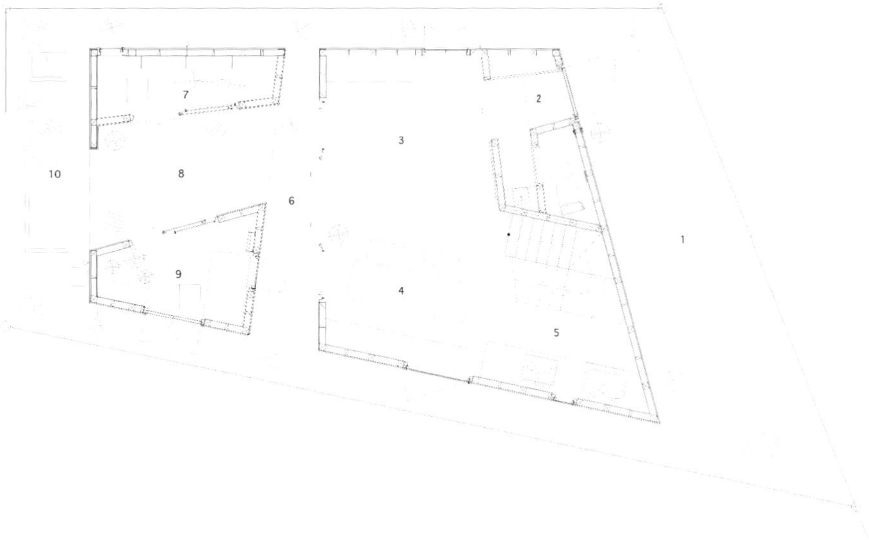

First Floor Plan/一层平面图

1:north garden
2:entrance
3:living
4:dining
5:kitchen
6:alley garden
7:Bicycle maintenance room
8:outer room
9:Atelier room
10:south garden
11:room
12:powder room
13:bath
14: bed room
15:clothet
16:hanmoc terrace
17:void
18:flower room
19:roof terrace
20:roof

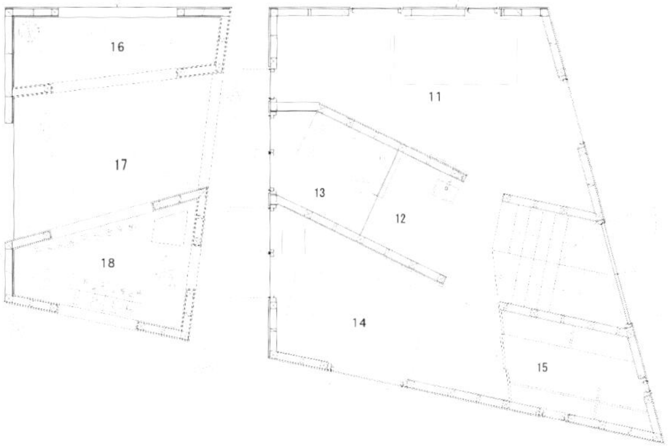

Second Floor Plan/二层平面图

1:north garden
2:entrance
3:living
4:dining
5:kitchen
6:alley garden
7:Bicycle maintenance room
8:outer room
9:Atelier room
10:south garden
11:room
12:powder room
13:bath
14: bed room
15:clothet
16:hanmoc terrace
17:void
18:flower room
19:roof terrace
20:roof

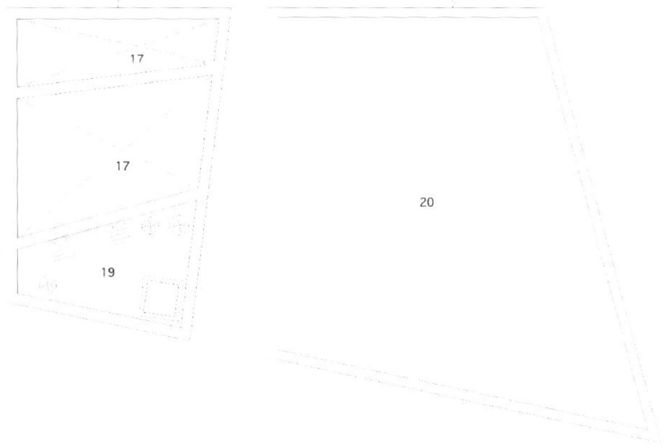

Roof Floor Plan/屋顶平面图

Wrap House

Future Studio

■ Architecture Design/建筑设计: Future Studio ■ Project Architect/项目设计师: Bunzo Ogawa ■ Location/地点: Hiroshima, Japan ■ Area/面积: site 108.75 m^2/building 64.46 m^2/total floor 118.86 m^2
■ Completion/建成时间: 2009 ■ Photograph/摄影: Toshiyuki Yano

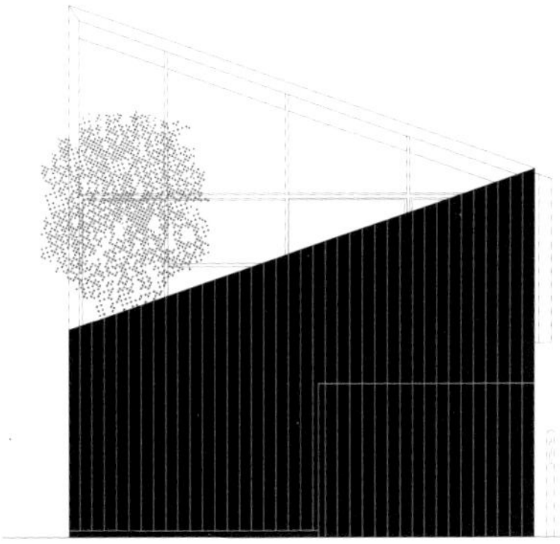

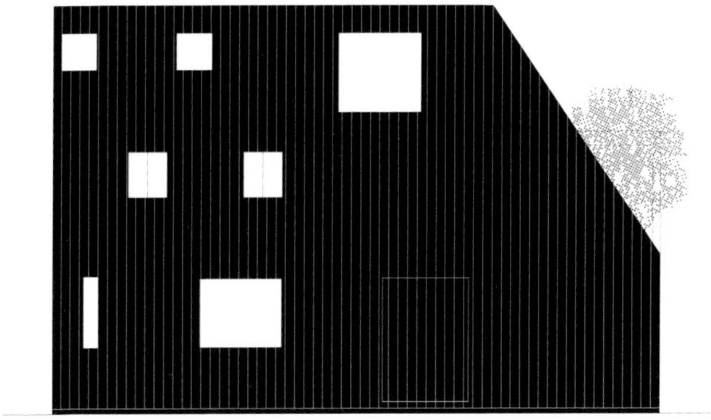

'WRAP HOUSE' is grouping and wrapping the three elements sunlight, spacious sense and privacy to enrich the living space.
The site is located in a flat residential quarter in the Southwest of Hiroshima City. As the site is from all sides surrounded by houses, one of which is directly bordering the site boundary, it was necessary to come up with the following architectural arrangement.
To create a bright and wide space while protecting one's privacy despite the site's restrictions, a wall wraps around the full site creating a 'void' on the north side while respecting the permitted floor area ratio. The north orientated living space is extended by a terrace, which is located adjacent to the parking space while facing the street to the north. In addition, considering the natural sunlight conditions for the neighboring house to the west, a pitched roof was chosen and a v-shaped wall is surrounding the 'void' on the east side. As a result, sunlight shines into the void in the morning and a 'sunlight well' is created. The collected sunlight is reflected by a white wall, and illuminates the room gently. In the early morning, sunlight enters the void from the east. As the sun rises further it shines through the upper window, and when the sun begins to decline light falls through the skylights.
When the night comes, the diagonally cut wall appears while the scenery around it disappears. A space different to the daytime emerges.
The external void space is perceived as an extension of the internal space, while the wrap-around wall is both protecting the privacy of the residence space against the surroundings and creating a spacious impression. The sky can be seen from the bathroom on first floor, while the car can be overlooked from the master bedroom.
The entrance for both residents and car is facing the road to the north and is formed by a part of the wall which can be opened and shut providing access to the void space. The sky framed by a diagonal wall extends to the top and an ash tree with a height of 6m penetrates the second floor terrace. This north side buffer zone, although not included in the legal building footprint, plays a role in maximizing the site's potential.
"WRAP HOUSE" with its north facing void is a proposal of a new architectural style resulting from restrictions.

"包裹的房屋"将三个基本元素——阳光、空间感、隐秘性集合并加以包装，以达到丰富居住空间的目的。
该项目位于广岛市西南的一片平坦的居住区内。由于该地块的周围是鳞次栉比的房屋，而其中的一栋建筑更是紧邻该址，所以，设计师提出以下的建筑设计方案，是很有必要的。
在充分保证个人隐私的情况下，设计师克服了该地不利的条件，建造了一堵高墙，由此，在北面创建出一个"中空结构"。朝北的起居空间被车库旁的一个面朝街道的平台大大拓展了。另外，考虑到西侧邻居住宅的采光情况，设计师选用了一个倾斜的屋顶和一面V形的墙，将东侧的"中空结构"包裹起来。因此，清晨的阳光可以照进中空结构，提升采光度。而收集来的阳光经白墙反射，又柔和地照亮了整个房间。每天清晨，阳光从中空结构的东侧照射进来，当太阳逐渐升高时，阳光就会透过上层的窗户洒向室内，而当太阳逐渐下落时，阳光又可以通过天窗照射进来。
当夜幕降临的时候，那面斜切墙会在周围景象消隐的同时显现出来，形成一种与白天不同的空间效果。
外部的中空结构，可被视作内部空间的延伸，而包裹的墙体既保护了居住区的隐秘性，又营造出宽敞的感觉。从一楼的卫生间里，能够眺望到天空；从主卧里，可以看到车库的情况，非常实用。
人和车的入口朝向北面的公路，由墙体的一部分构成，可以开启，当墙体关闭时，可以由此进入中空区。在横切的墙体上部，可以看到天空的一角，一棵高为6米的白蜡树则穿过了二层的平台。该建筑北侧的缓冲区，尽管不属于正规设计的范畴，但仍在很大程度上挖掘了该地的潜能。
该"包裹的房屋"，以其北面中空的结构特性，为那些在种种限制条件下进行的设计，提供了一种新的尝试。

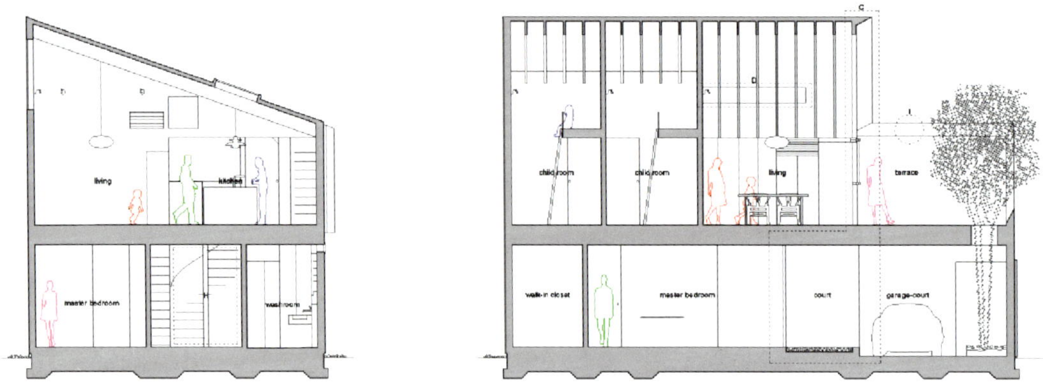
living
kitchen
master bedroom
washroom
child room
child room
living
terrace
walk-in closet
master bedroom
court
garage-court

Section/剖面图

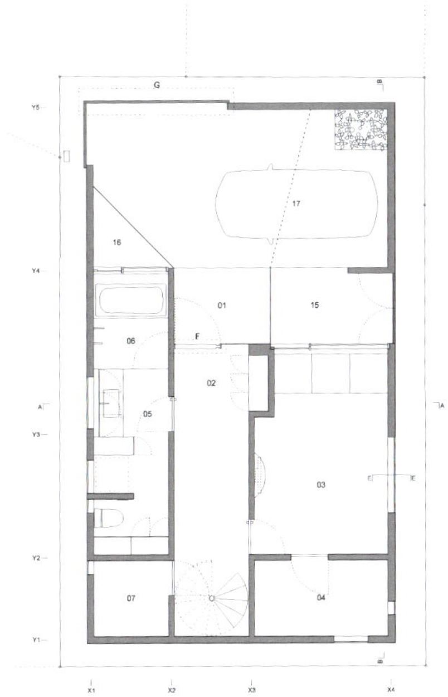

01.porch
02.entrance hall
03.master bedroom
04.walk-in closet
05.washroom
06.bathroom
07.storage
08.kitchen
09.dining
10.living
11.child room
12.loft
13.toilet
14.terrace
15.court
16.bath-court
17.garage-court

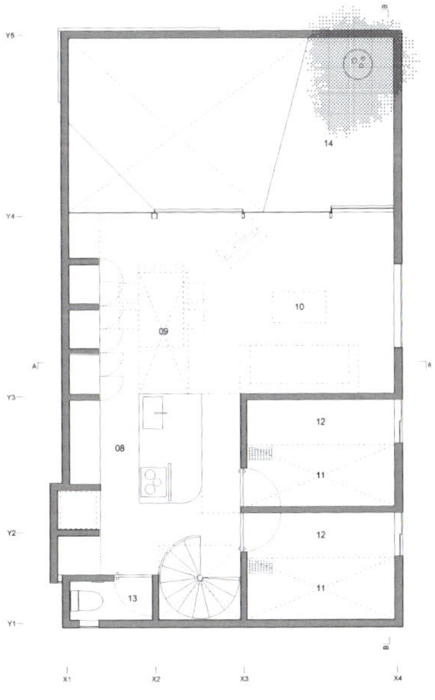

Plan/平面图

Orchid House

Andrés Remy Arquitectos

■ Architecture Design/建筑设计: Andrés Remy Arquitectos ■ Desing Team/设计团队: Andrés Remy, Hernán Pardillos (asociated), Paula Mancini, Lucila Lopez, Julieta Rafel
■ Collaborator/合作者: Lilian Kandus, Coral Banegas, Gisela Colombo, Diego Siddi, Martin Delatorre, Juan Etala ■ Loation/地点: Buenos Aires, Argentina ■ Area/面积: total 465 m²
■ Completion/建成时间: 2008 ■ Photograph/摄影: Alejandro Peral

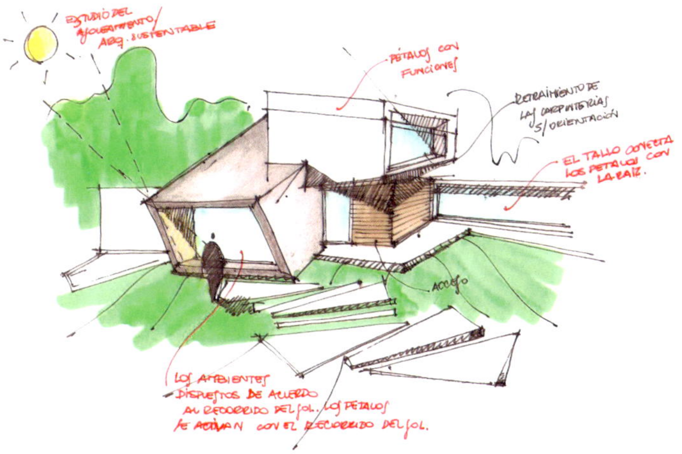

The Orchid was a challenging project, committed by a young couple with two sons, enthusiastic about sustainable architecture. After consulting several architects, they gave the commission to Andrés Remy, who had investigated these concepts in New York at Rafael Viñoly's Studio for four years.

Sustainability implies a lot of varieties, such as efficient and rational use of energy and water, natural ventilation and lighting, and low-environment-impact materials.

The house has the best orientation, possible thanks to the large lot.

The concept came from the client's hobby, growing orchids. The house is based on the different parts of the orchid: the roots, the stem and the flower.

The sun rays impact in the interior of each room, were also studied, to determine the optimal depth to place the windows. This gives a unique volumetric outcome to the project. Taking advantage of sun rays in winter increases the interior temperature up to a comfortable level. The design includes glazed volumes with good thermal insulation and small windows in the worst orientation, such as the south facade.

The windows consist of aluminum frames supplied with thermal bridge breaker and double-hermetic-glass. A wide variety of insulating materials were also used, as well as water based paint and wall and roof air chambers.

In the lower floor, the convenient location of the opening windows allows natural air flow helping to decrease the humidity in the room. What's more, the big thresholds create a good distribution of fresh air.

All this, gives account of the complexity of the house, a glass and concrete flower, designed according to a program of needs and the client's concerns.

本项目是应一对热衷于环保建筑的年轻夫妻和他们的两个儿子之托而设计的，富有挑战性。他们在向多名设计师咨询后，将这个设计任务委托给了曾在纽约的Rafael Viñoly工作室工作四年的专门研究过这些理念的设计师Andrés Remy。

可持续性包含很多方面,比如高效合理地利用能源和水资源，建筑自然通风、采光情况良好以及使用低污染材料。

因为有足够宽敞的场地，这个房子朝向很好。

建筑的设计理念源于客户平时的业余爱好——种植兰花。而这栋房子的设计灵感正是源于兰花的根、茎和花等不同部分。

光线能够照进每一个房间，就是按照经过专门设计、通过测定得出的最佳照射长度，来设计窗户位置的。该设计给这个项目带来了意想不到的效果。在冬日里，建筑能充分利用阳光照射，使室内温度适宜。该设计包括采用了良好的保温材料涂层，并在采光不佳的房间，设计使用了很小的窗户。

这个窗户是由铝合金窗框和带有热桥的双层密封玻璃构成的。当然，同时采用了很多种类的保温材料和水溶性涂料以及墙面和屋顶的保温层。

在楼下，方便的开放式窗户能够保证通风换气，有助于减少房间内的潮气。

所有的这些，都是根据项目的需要和客户的想法进行的，从而把这栋房子建成了一座兰花造型的、纵横交错的玻璃和混凝土结构小屋。

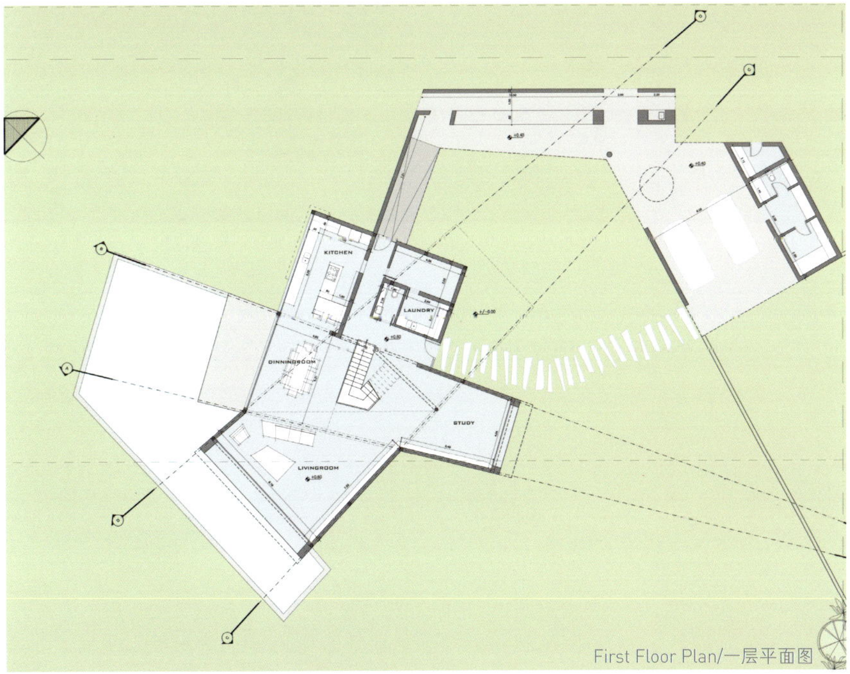

First Floor Plan/一层平面图

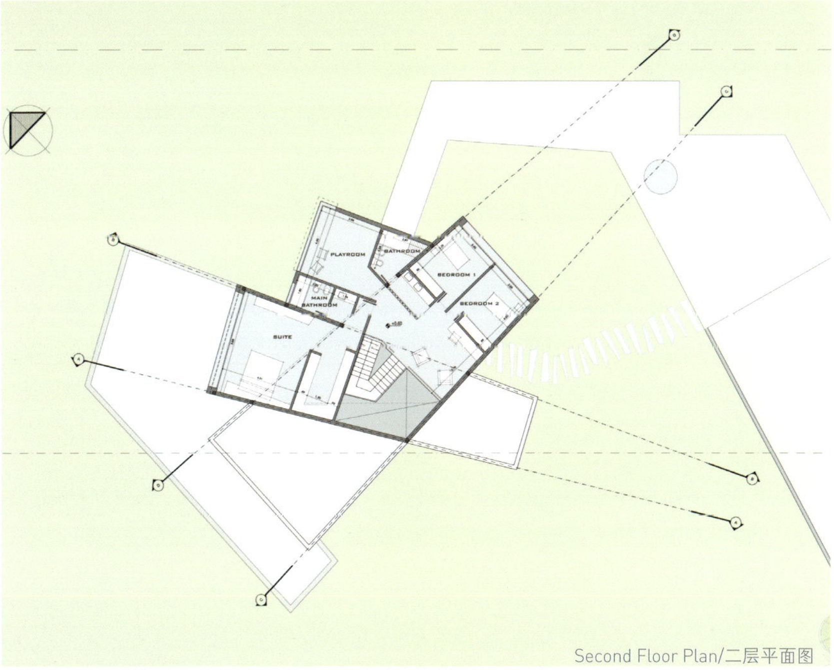

Second Floor Plan/二层平面图

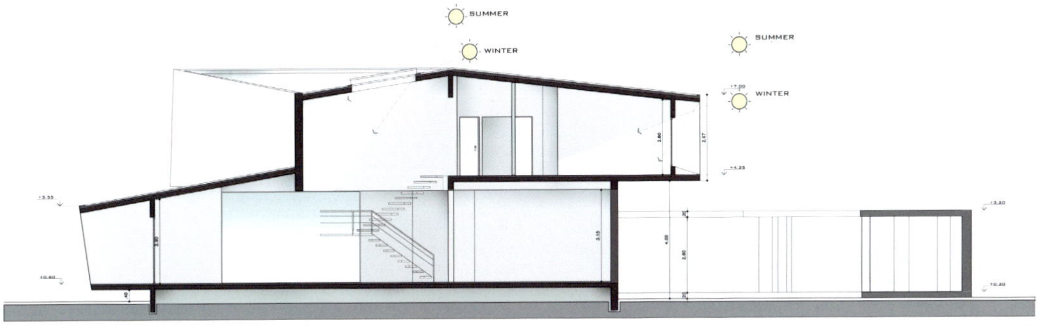

Section/剖面图

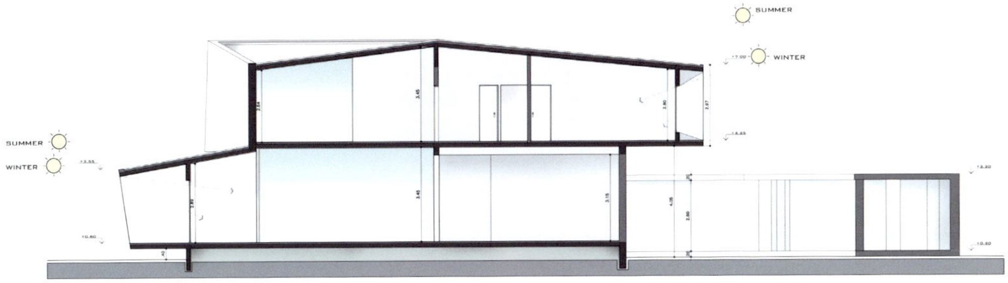

Section/剖面图

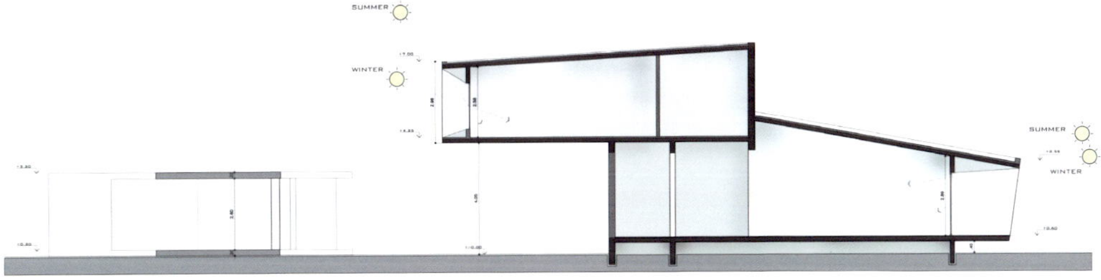

Section/剖面图

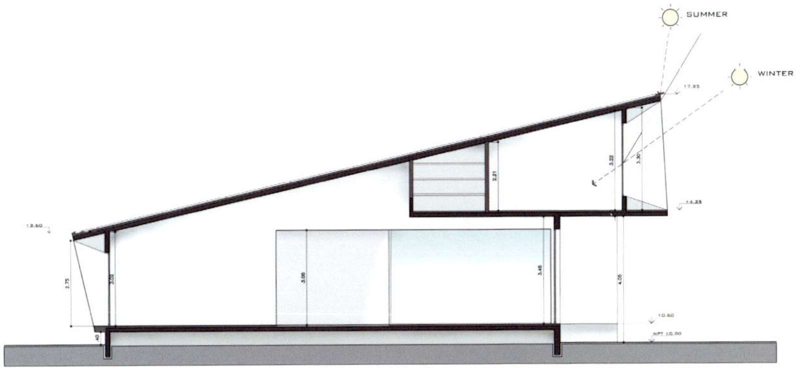

Section/剖面图

Devoto House

Andrés Remy Arquitectos

- Architecture Design/建筑设计: Andrés Remy Arquitectos
- Design Team/设计团队: Andrés Remy, Carlos Arellano, Hernán Pardillos (asociated), Julieta Rafel, Lilian Kandus, Diego Siddi, Martin Delatorre
- Area/面积: 550 m^2
- Completion/建成时间: 2009
- Photograph/摄影: Alejandro Peral

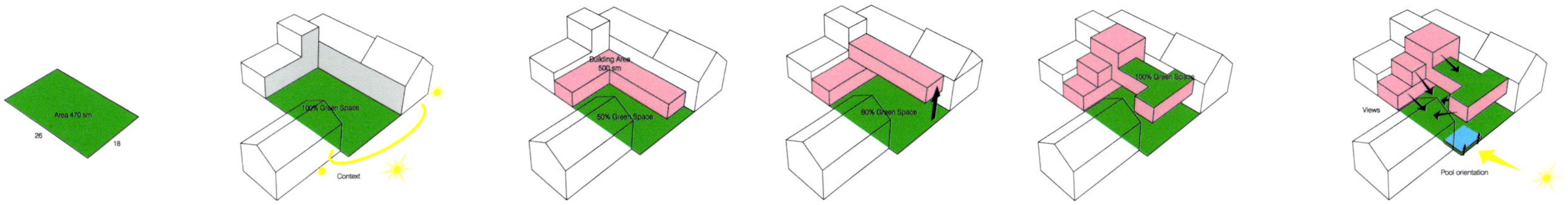

This project was a great challenge for the Studio, considering that it was going to take part of the urban structure. The lot of 18 x 24 meters is located in Devoto. It is between two existing buildings that threatens the project when the problem of isolation and view.

For this reason this project is leaning towards its neighbor, creating an introspective posture that creates the views from the inside rather than from the outside. An architecture that generated its own views.

The house of 500 m² was done for a young couple with kids. It was larger than the dimensions of given lot, threatening the space of the garden. For that reason, the first floor was given less program, leaving open the exterior spaces.

The living, kitchen and dining rooms expose the garden that penetrates all the volumes in the house. The stone wall guides the inhabitant within the house, therefore, the rough texture of the streets are soften by the water that flows down the stone wall. The green area between the living room and the neighbor guides the way to the entrance of the house.

In this project, the impact of the sun path was carefully studied, especially to place the swimming pool. The swimming pool is elevated from the ground creating a glass wall, creating views from different points in the house and focal point outside the house.

The house is formed vertically having a shifting effect. The bedrooms are located in the east overlooking the heart of the house, the garden. The last floor places a spa; having a jacuzzi, sauna and a gym with an exterior space. Privately settled for the couple.

This Project goes beyond the rigid setting. It translates the urban setting into a natural setting subtlety. It pushes the boundaries of the conventional urban home to a spacious living where one has the best of both, the natural and the urban setting.

因为这栋房子即将成为城市景观的一部分，所以该项目对工作室来说是个巨大的挑战。这块地面积是18米×24米，位于Devoto地区。它夹在两个已有建筑物之间，这样对房子的独立性和视野造成了一定的影响。为此，该项目向它的临近建筑倾斜，构造出一个内翻式的造型，这样就从里面而不是外面，营造了可视景观。可以说，每一个建筑物的设计都有它独到之处。

这栋面积为500平方米的房子是为一对年轻夫妇和他们的孩子建造的。实际使用面积大于原规格尺寸，这对花园多少有些影响。考虑到这一点，设计师没有对第一层做过多设计，而是让其面向外部空间。

客厅、厨房和餐厅面朝花园，这样就可以把花园的一切尽收眼底。从石墙流下的细水柔化了街巷的粗犷感。客厅和邻居之间的绿色区域指示着通往房子入口的路径。

在该项目中，设计师充分考虑了阳光的作用，尤其考虑到阳光对游泳池的影响。这个游泳池是一个由玻璃壁建成的地上泳池，不仅为室内的不同角度营造出不同景观，而且成为室外的一大亮点。

该建筑物从垂直的角度来看有一种移位感。卧室位于东侧，人们能够在此俯视房子的中心花园。房子的底层设置了一个做SPA的地方，可以享受到水流按摩、蒸桑拿，另外室外还有一个健身房，这是为夫妻设计的。

该设计超越了僵硬的设置， 它将城市化布局经过稍微转换后变得如此亲近自然，把传统城市住宅的界限放宽了很多，由此，一栋房子可以既城市化又贴近自然，一举两得。

Plan/平面图

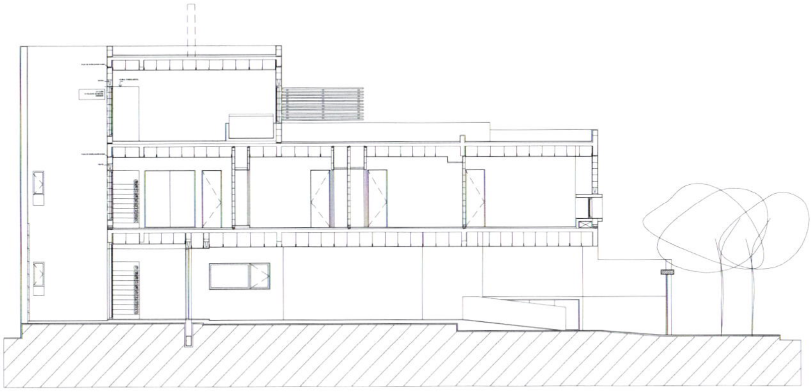

Section/剖面图

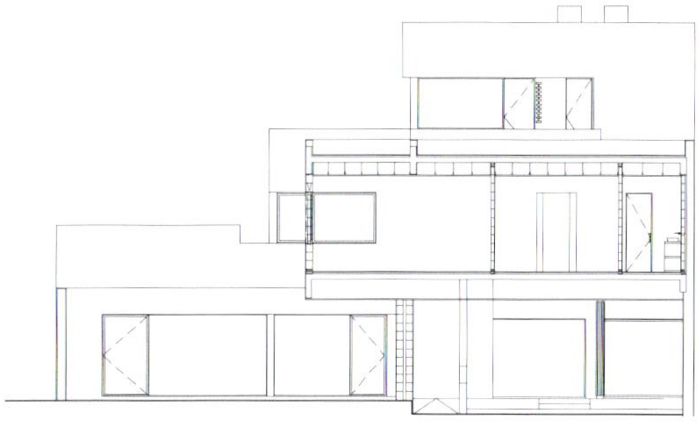

Section/剖面图

Sapphire Gallery

XTEN Architecture

■ Architecture Design/建筑设计: XTEN Architecture
■ Project Architect/项目建筑师: Austin Kelly, Monika Haefelfinger
■ Location/地点: California, USA
■ Area/面积: 139.4 m²
■ Completion/建成时间: 2009
■ Photograph/摄影: Art Gray

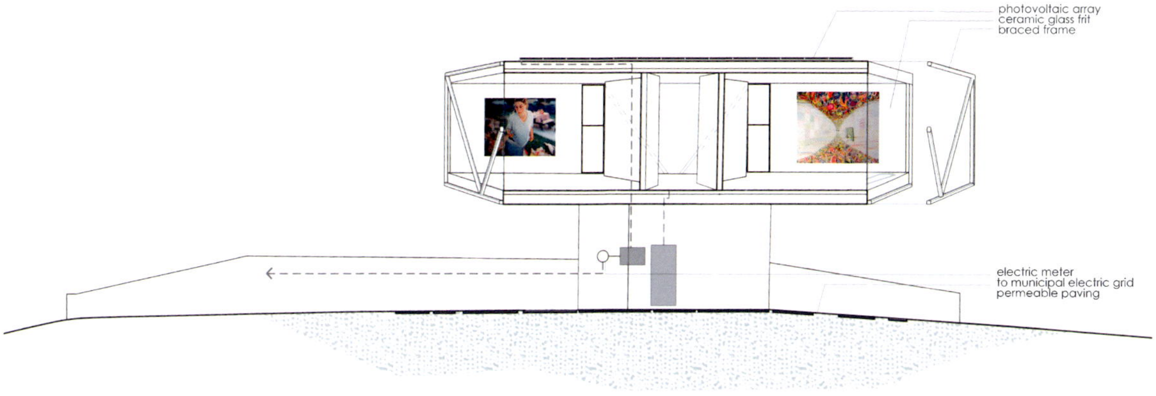

The Sapphire Gallery is a residential addition designed to display a private collection of contemporary art while also providing for a home office with views to the surrounding hills. The owners' collection includes work by the artists Gregory Crewdson, Uta Barth, Tomoroy Dodge and the video artist Jennifer Steinkamp, and they wanted a new building that would be more than just a container for their art collection. The new gallery extension is multivalent, with different spaces for presenting artworks, while also opening to hillside views, incorporating a home office, and creating a compelling new focal point for the approach and entry to the property.

The new structure is grafted onto the circulation spine of the existing house and lifted off the ground to provide a minimal footprint. Freeing the ground plane creates a new hardscape/landscape area for the family that they use as carport, children's play area, for art parties and video projections.

A structural system of lightweight braced frames was developed to achieve the double cantilevers at each end of the trapezoidal building, which were factory built and assembled by crane in one day. These trusses rest upon moment frames that clear span the open ground plane in the perpendicular direction, allowing the floor and roof diaphragms to be framed conventionally. The system proved to be a remarkably simple, flexible and cost-effective way to achieve the program parameters of the project.

The remaining details of dark quartz pebble flooring, white steel stairs and perforated steel railings, full height pivot doors, and walls of UV treated glass are designed as a quiet backdrop to the artworks and natural surroundings. An array of photovoltaic cells on the South facing sloped roof produces an average of 15kWh per day, enough to supply all the energy for the new building with a surplus directed towards the main house.

Sapphire 画廊是一个私人的当代艺术品收藏馆，同时主人们可以在此居住，享受周围山丘的风光。主人的藏品包括了艺术家Gregory Crewdson、Uta Barth、Tomoroy Dodge和录影家Jennifer Steinkamp的作品，他们希望这里不仅仅是一个存放藏品的场所，而是能够包含更多意义。为了在不同的空间展示作品，也为了能够充分地享有这片山丘景色，设计师加入家庭办公室的元素，这也让来到这里的人眼前一亮。

新的房屋结构是在原有结构的基础上改建的，设计抬高了原来的底层，腾出了一个小空间。开辟的这么一小块地方，可以用作家庭的车库或供孩子们玩耍，也可以举办艺术聚会或者是投射影像。

设计师开发了一种轻型框架支撑结构体系，这种材质在工厂从加工到组装只需要一天的时间。这些构架依托精准的框架在垂直方向横跨开放的地平面，楼板和屋顶被分别框起来。这个体系在完成该项目的过程中起到了决定性作用，凸显了简约、灵活且经济划算的特点。

还有其他细节，黑色的石英卵石地板、白色的钢制楼梯、穿孔钢围栏、通顶的枢纽门和抵御紫外线的落地玻璃，这一切的设计都是为那些艺术品和自然景观做陪衬。一个光伏电池阵列放置在朝南的斜坡屋顶上，每天能够产生15千瓦时的电量，足够供应这栋新建筑的能耗，而其剩余电量则供主体建筑使用。

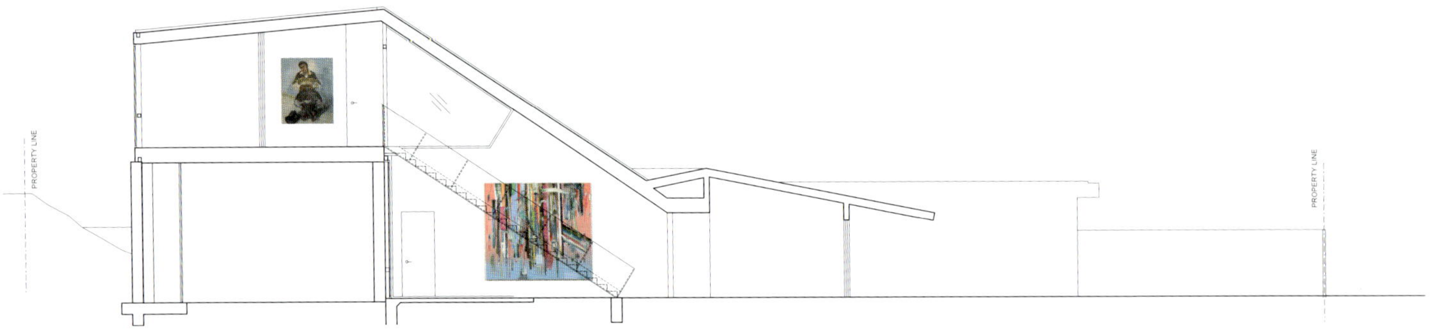

Section/平面图

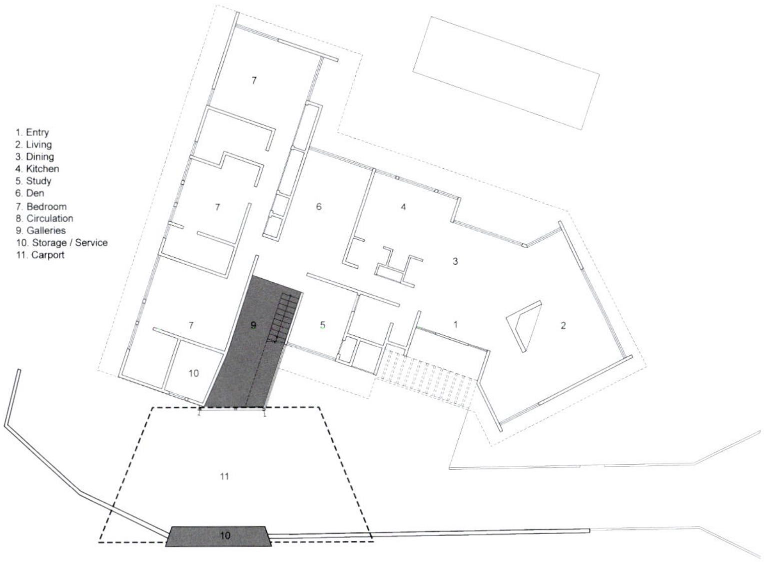

Plan/平面图

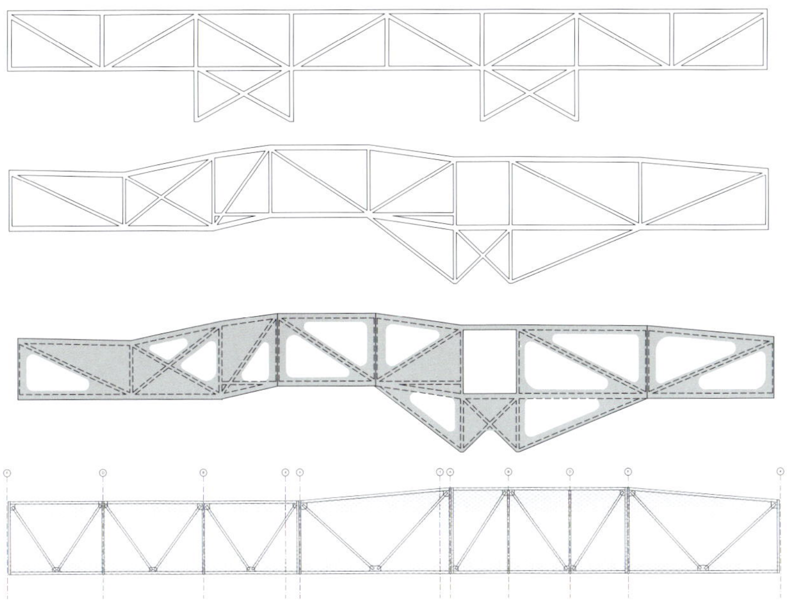

Unfolded Structure/展开结构图

Diamond House

XTEN Architecture

- Architecture Design/建筑设计: XTEN Architecture
- Project Architect/项目建筑师: Monika Haefelfinger, Austin Kelly, AIA, LEED AP
- Client/委托方: Aisha Ayers
- Location/地点: California, USA
- Area/面积: exterior 111.5 m^2
- Completion/建成时间: 2009
- Photograph/摄影: Art Gray

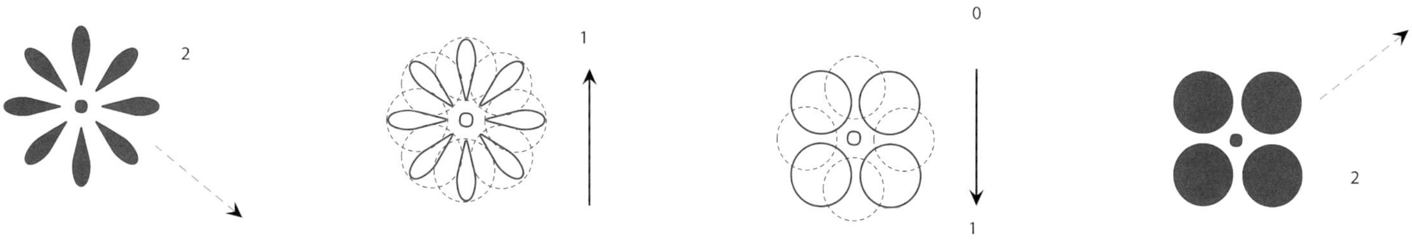

The Diamond House is a sound studio and office extension to a house located deep in a canyon, against a severely sloping hillside, with minimal access and little space upon which to build. Direct sunlight reaches the site for only a few hours a day. The geotechnical condition is challenging, requiring 30-foot caissons to underpin new walls and foundations. A complex web of regulations governed the height, width, depth and specific relationship to the retaining walls needed to build the project.
Given these constraints, a multifaceted architectural strategy was developed for the small building. First, a base building geometry was developed to conform to the hillside and required codes while maximizing the interior spaces by extending them into adjacent sideyards. Like a rock placed in a small pond, the addition is carefully placed between the existing structure and an imposing hillside to inflect the landscape and create exterior programmatic spaces around it where none could exist before. The building geometry also conforms to the interior program as a corner of the upper floor flares out to accommodate a writing desk built into a north-facing window and a series of wall planes fold up and over the building to create a rooftop railing and enclosure.
Next, a building material system was developed to both relate the new extension to the natural landscape and to reduce the visual scale of the building. The facade pattern was created from natural elements taken from the canyon site; abstracted, scaled and arrayed across the building in 3d modeling programs.
Many patternsand scales were investigated in this manner, from which several full size prototypes were made by laser cutting and cnc routing aluminum, zinc, steel, plywood and cement panel samples.

Diamond House项目是一个录音室，同时也是家庭办公室。地处峡谷深处、陡峭的山坡前，狭小的通道与空间里，建筑师建造了这个建筑物，这栋房子每天只有几个小时的日光直射。陡壁峡口的地理条件非常差，为了巩固新的墙面和地基，建筑师需要在9.1米的地下作业。为了控制建筑高度、宽度、深度以及与挡土墙的具体关系，建筑师运用了一整套复杂的规程。

建筑师为这栋小房子构想了一个多层面的建筑思路。首先，根据紧邻的山坡和所需的规格设计了建筑的基本几何图形，并最大限度地利用室内空间，使其扩大到邻边的侧院。该设计仿佛是将一块巨石放在水池中，小心地建在现有建筑和庄严峭壁之间，使其室外空间环绕着前所未有的秀丽风景。根据建筑的几何图形，在楼上的角落也延展出一块空间用来放置一个写字台。而几面可折叠的墙面在一起组成了一个屋顶围栏，从而形成一块单独的空间。
其次，建筑材料体系的选择，需要考虑接近自然景观的新型延展性材料和降低建筑物视觉范围的材料。建筑表面的设计源于峡谷本身的自然元素，是抽象的，且整个建筑的规模和排列都是三维模式。

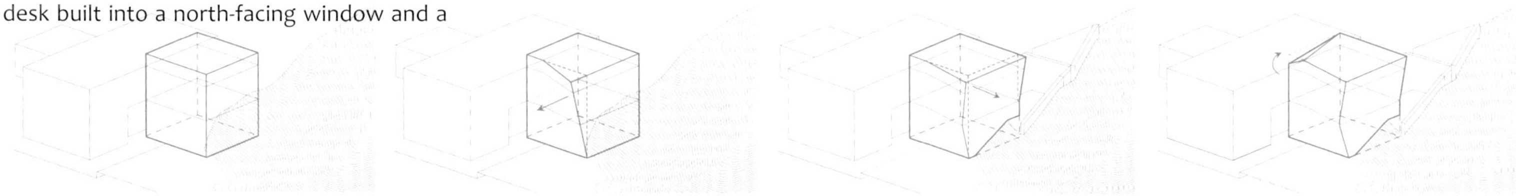
(1) Initial Site Condition
(2) Programmatic deformation
(3) Roof terrace deformation
(4) Fold / delamination over from existing

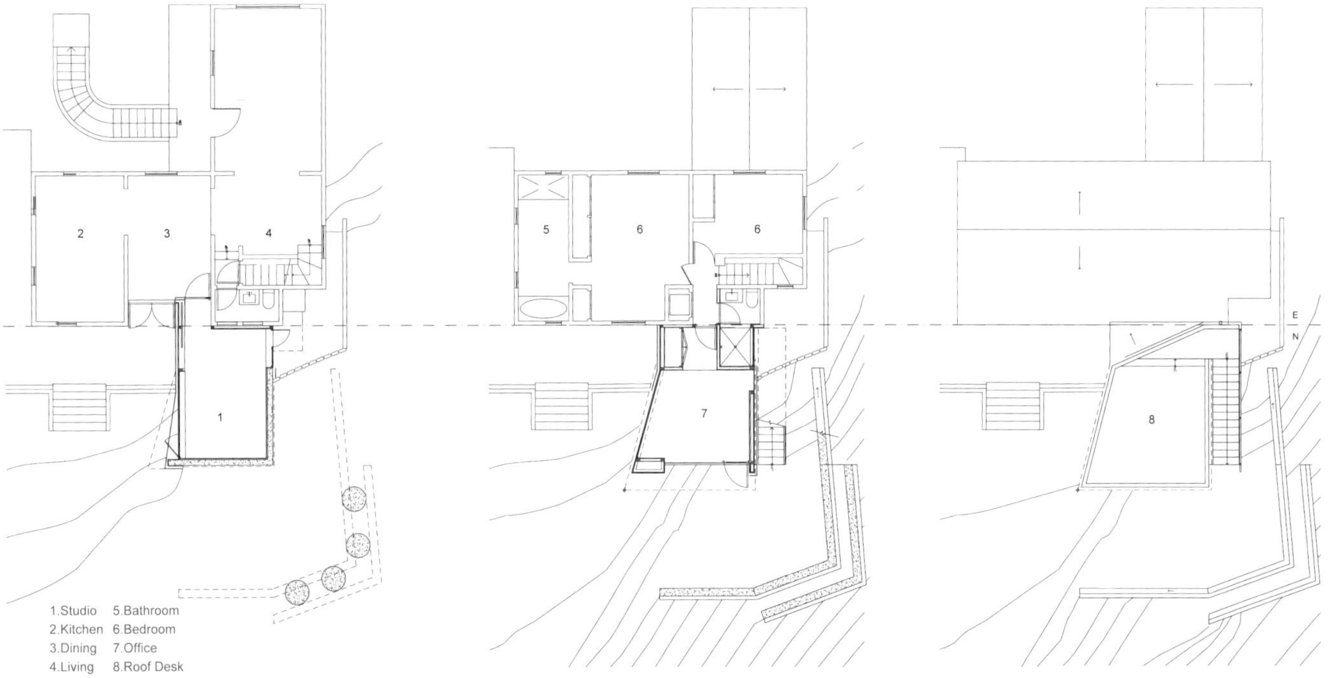

Plan/平面图

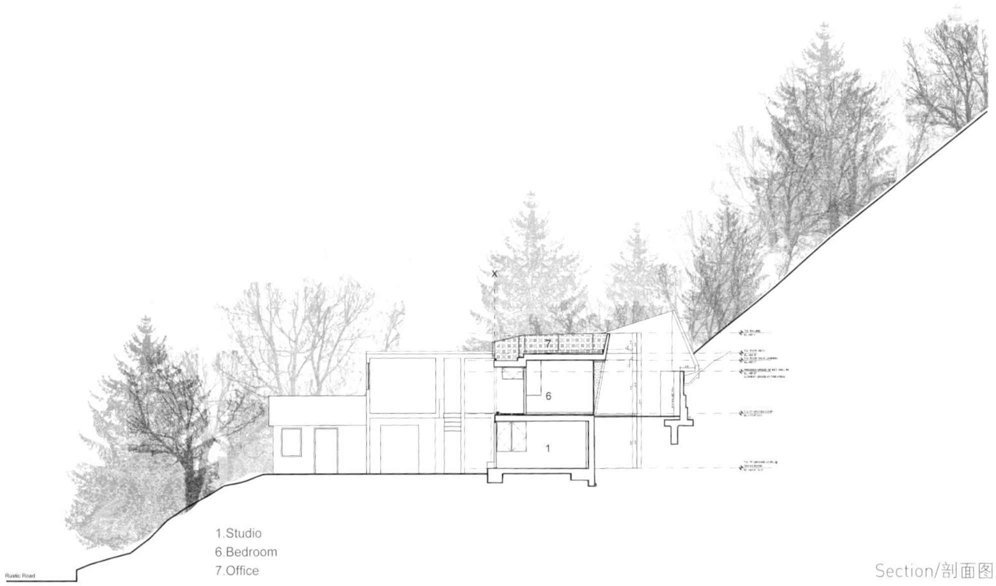

Section/剖面图

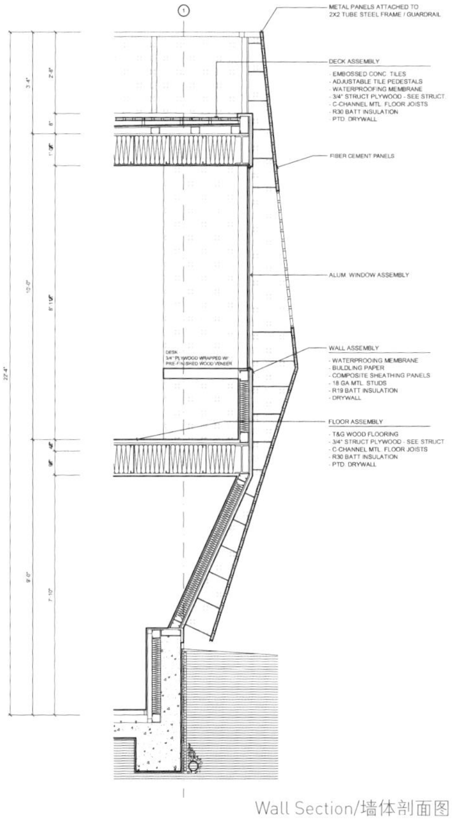

Wall Section/墙体剖面图

VilLA NM

UNStudio

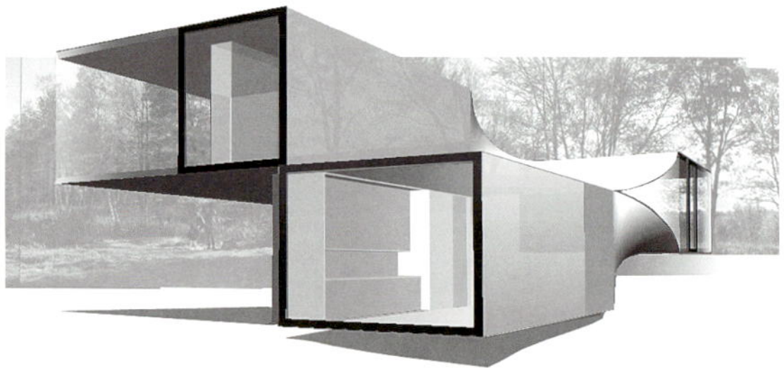

■ Architecture Design/建筑设计: UNStudio ■ Design team/设计团队: Ben van Berkel with Olaf Gipser and Andrew Benn, Colette Parras, Jacco van Wengerden, Maria Eugenia Diaz, Jan Debelius, Martin Kuitert, Pablo Rica, Olga Vazquez-Ruano ■ Location/地点: Upstate New York, USA ■ Area/面积: gross floor surface 250 m² ■ Photograph/摄影: Christian Richters

structure of volume

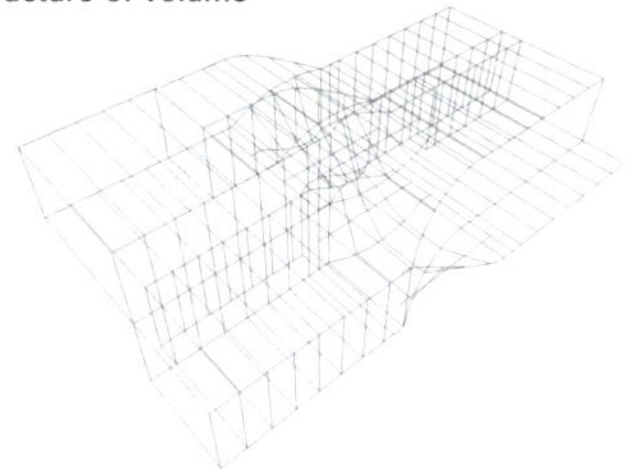

planar volumes

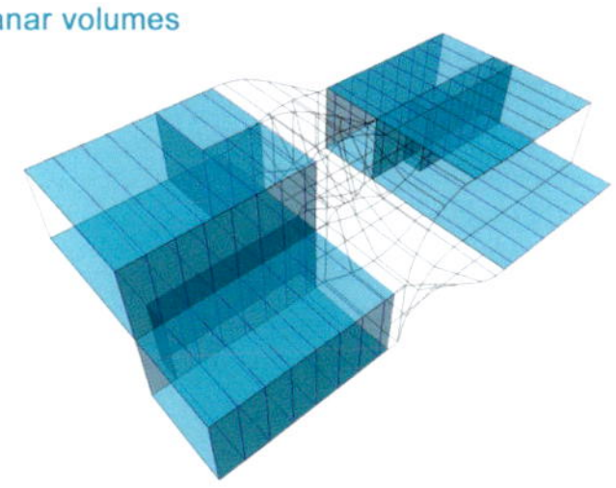

twisted surfaces

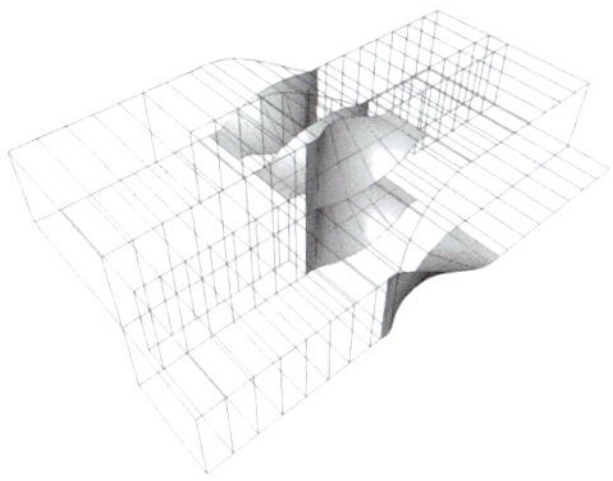

single-curved surfaces

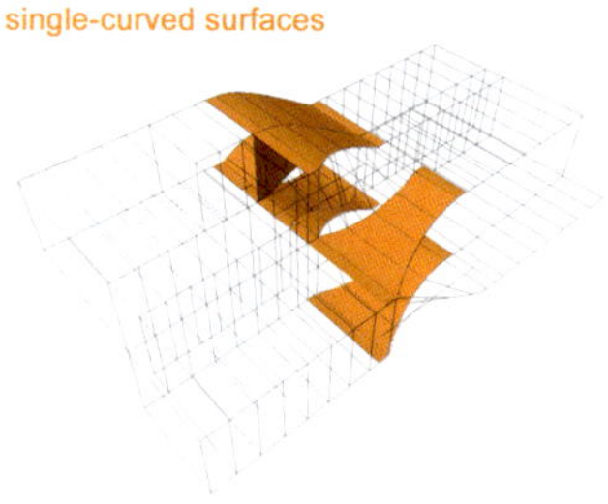

VilLA NM is a family summerhouse situated in a hilly and woody area, some two hours drive from NYC.

The house is situated on a plot of 7,000m^2 with a tremendous 360° view of the forests and meadows. The sloping site is used as device for the programmatic and volumetric organization of the house. A single box-like volume bifurcates into two separate volumes; one seamlessly following the Northern slope; the other lifted above the hill creating a covered parking and generating a split-level internal organization.

The volumetric transition is generated by a set of five parallel walls that rotate along a horizontal axis from vertical to horizontal. The walls become floor and vice versa. The ruled surface maintaining this transition is repeated five times in the building. The interior space also takes advantage of the split-level organization. The kitchen and dining area on the ground floor are connected by a ramp to the living space above, the 1.5-meter (5 feet) height change allowing for a tremendous view over the valley. A similar ramp connects the living area to the master and the children's bedrooms on the second floor.

All the functional facilities such as the bathroom, kitchen and fireplace are situated in the vertical axis of the house. Rooms that require a higher level of privacy are partly closed of to the exterior. All other rooms are provided with large glassed windows.

The materialization of the design is a combination of concrete and glass with a light metal construction. The intelligent transformation of a geometric form and the standardization of the structural elements enabled the economical production of a highly individual building with strong spatial qualities.

VilLA NM项目是一个家庭暑期度假村，处在一个树木茂盛的丘陵地带，距离纽约城两小时的车程。

这栋房子占地面积约为7 200平方米，拥有全方位的视野，能够将周围的森林和草地尽收眼底。斜坡地段用来规划房子，扩展现有空间。一个单一盒状的空间被分为两个单独的部分，一部分紧贴北部山坡，而另一部分则在丘陵之上建起了一个室内停车场，并产生了一个错层的内部结构。

室内空间是由一组五排平行的墙壁沿着从垂直到水平的水平轴旋转而成的，这样墙壁变成了地板，地板变成了墙壁。这栋房子为保持该现状，按照直纹曲面的方式重复了五遍。内部空间还利用了错层的结构优势，使地下的厨房和餐厅相连，并通过一个坡道与上一层的客厅相连，由于室内高度被抬高了1.5米，因此住户能够对广阔山谷拥有一览无余的视野。一个类似的坡道从客厅通向二层主人和孩子们的卧室。

房屋的所有设施如浴室、厨房、壁炉都安置在房子的垂直轴面上。客房需要更多的隐私，因此房间对外都是半封闭式的，而其他的房间均安装有大面积玻璃窗。

设计材质是由玻璃和轻金属混合而成的。通过智能改造和结构元素的标准化处理，使其拥有强大的空间质量保证。

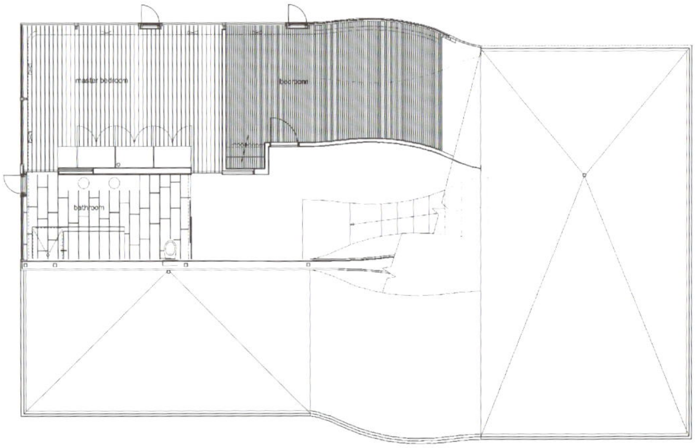

Second Floor Plan/二层平面图

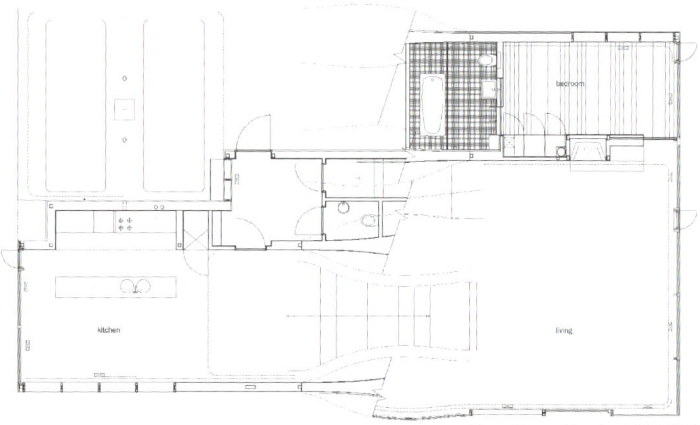

First Floor Plan/一层平面图

View From South-East/东南面效果图

View From West/西面效果图

combination of twists

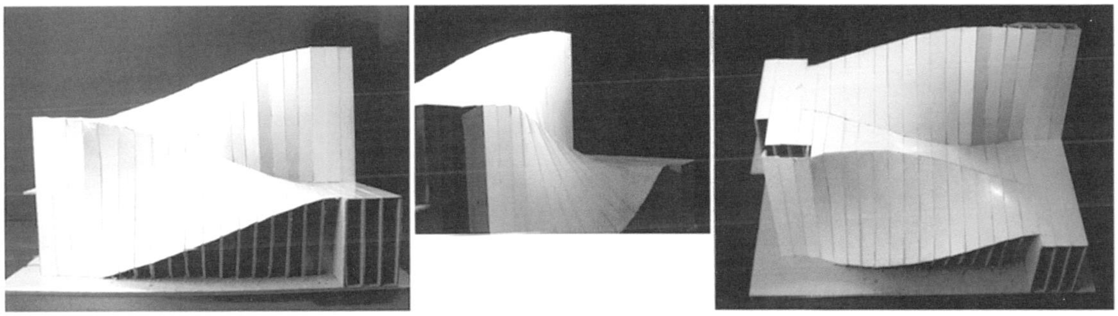

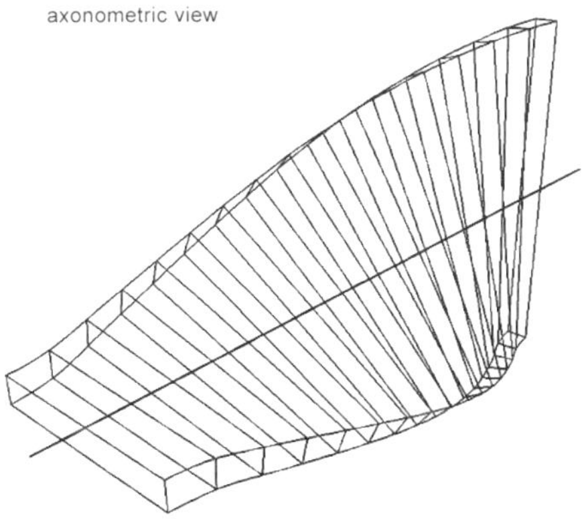

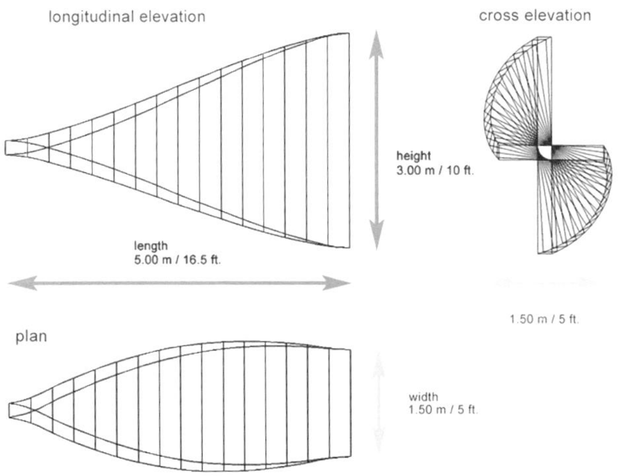

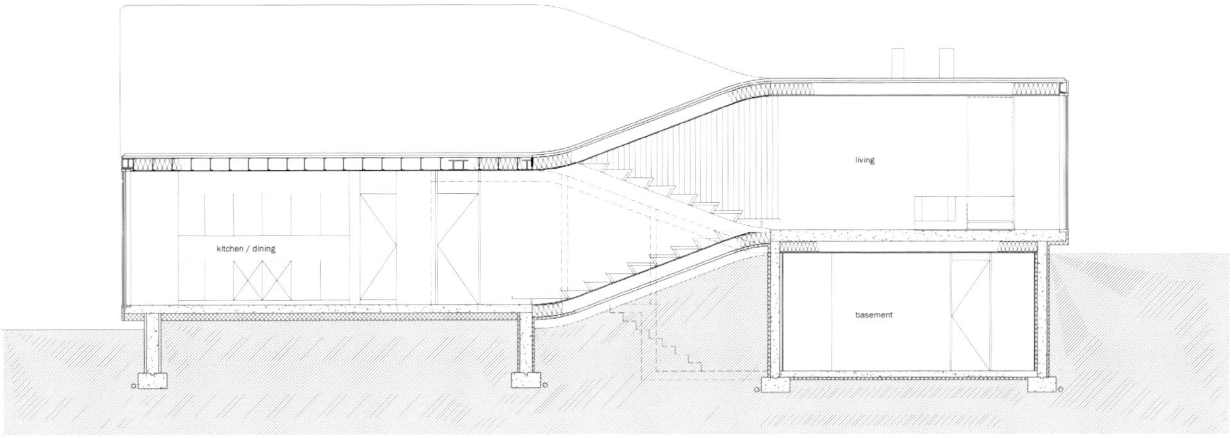

Section: West/西剖面图

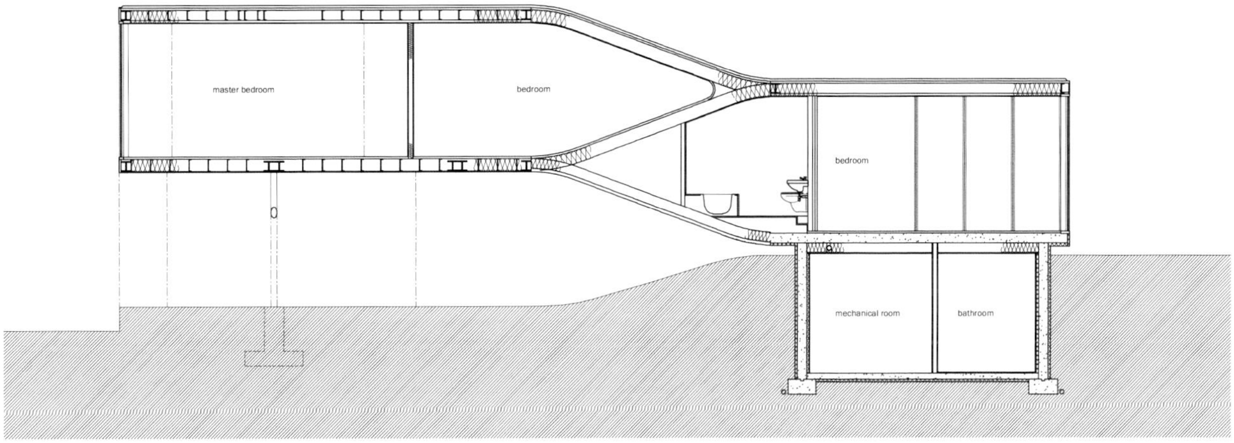

Section: East/东剖面图

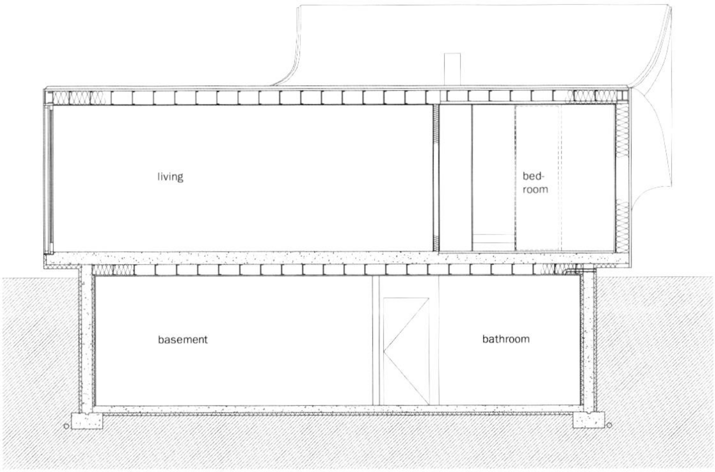

Section: North/北剖面图

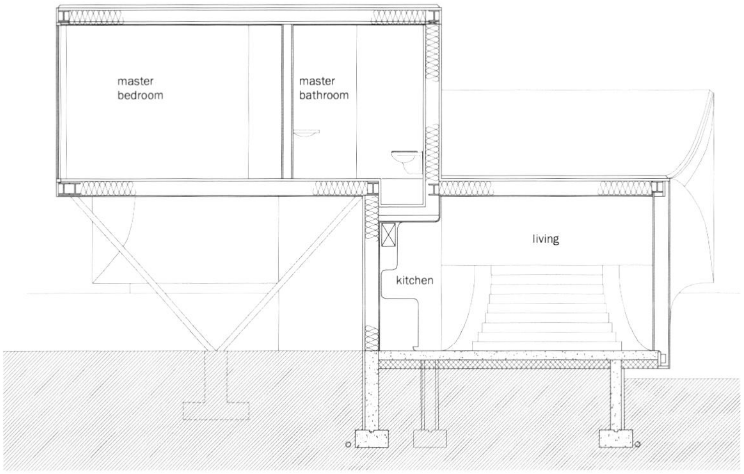

Section: South/南剖面图

Make-It-Right Duplex

Pugh + Scarpa Architects

■ Architecture Design/建筑设计: Pugh + Scarpa Architects ■ Project Architect/项目建筑师: Lawrence Scarpa, FAIA ■ Location/地点: New Orleans, Louisiana, USA ■ Area/面积: 176.5 m^2
■ Completion/建成时间: 2011 ■ Photograph/摄影: N/A, rendering by Pugh + Scarpa

The house accomplishes this with several significant design moves.

The front porch, facing the street, provides a gathering place for neighborhood residents and relatives, and is divided into distinctive sub-zones: a platform for outdoor cooking and access to the home, front steps with bleacher-like seating areas, and a platform just big enough for two rocking chairs to take in the life on the street. The inviting flames, porch and sheltering roof work together to make the home the social locus of the neighborhood. Rather than isolating the home from its Lower 9th Ward neighbors this connection is an attempt to strengthen the social network of the neighborhood.

The scale of the porch is purposefully oversized and grand. Topping out at 38 feet above grade, the structure is meant to stand tall and proud as a symbol of neighborhood and city pride. The verticality and focal strength of the home is emphasized by the presence of a 12-foot high outdoor "cook pit" integrated into the front porch, facing the street. The "cook pit" is made for outdoor grilling and barbecues, with open flames visible from the street at eye level, and a rotisserie enclosure accessed from the porch. A chimney extends from the hearth enclosure past the roofline, acting as the home's axis mundi and providing the armature for the required egress ladder for rooftop flood refuge. The porch also includes step seating to engage the street. Even though the house is five feet above the street, the porch reaches down to grade and makes a direct connection between the house and street connecting the families of the duplex directly to the street while simultaneously providing a welcoming symbol to the neighborhood.

设计师通过一些关键的设计举措，实现了设计的理念。

面向街道的前廊，为附近的居民和亲朋好友提供了一个聚会场所，而且该场所还有着明确的分区：一个平台，可用于室外烹饪，也可作为住宅的入口；灰白色的前门台阶，可以落座；一个能放下两把摇椅的平台，可近距离地感受市井生活。激情的火焰，长长的门廊以及巨大的遮阴顶，使这栋房子成为街坊邻里交流的好去处。通过设计师的这一大胆尝试，该建筑不但没有从街坊邻中孤立出来，反而加强了它与周围邻居间的社区联络。

设计师有意将门廊的尺寸扩大，该建筑高11.6米，成为邻里乃至该社区自豪的象征，展现了雄伟高昂的姿态。朝向街道的门廊上建有一个3.6米高的室外灶台，增强了该建筑的功能性。这个灶台是用来进行室外烧烤的，从街头望去就能望到烧烤的火光，从前廊上走过，还有一个烧烤区的围栏。烟囱从这里伸出去，远远高过屋顶，能在洪汛期间支撑起救生梯子将人员输送到屋顶。门廊还设计了阶梯式的座椅，以方便人们落座。尽管房子比街道高出1.5米，但其前廊逐渐延伸落到与街道水平的高度，使得该房子的主人一家可以直接与街道相连，由此可见，该建筑也象征了街坊邻里间的和睦相处。

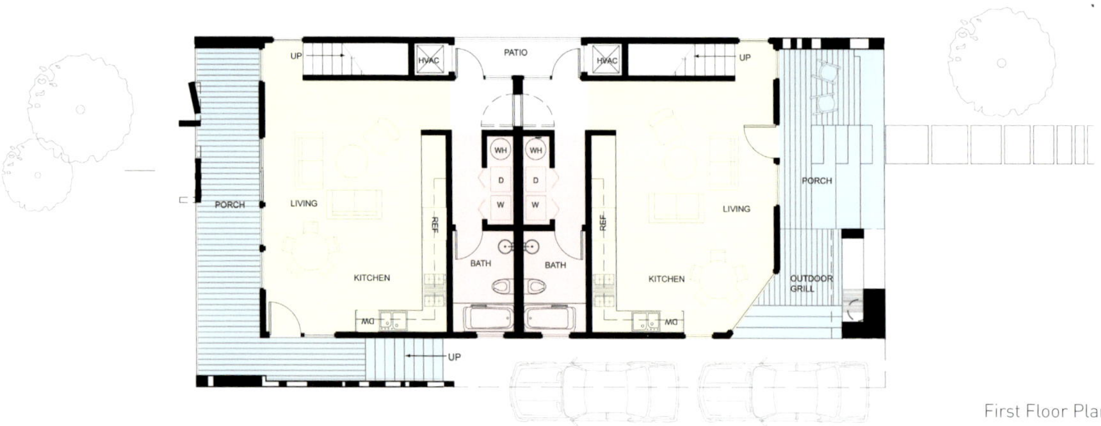

First Floor Plan/一层平面图

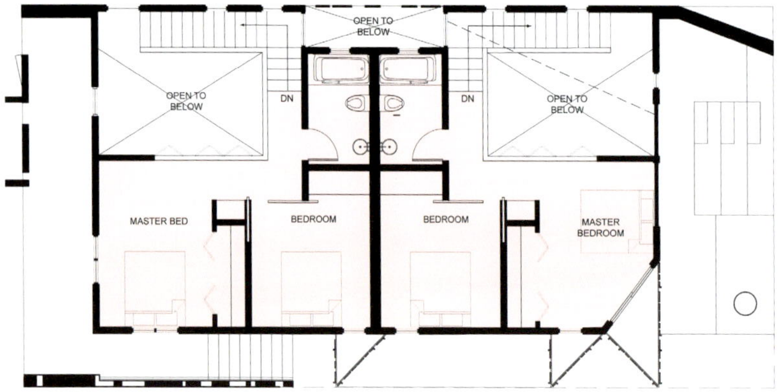

Second Floor Plan/二层平面图

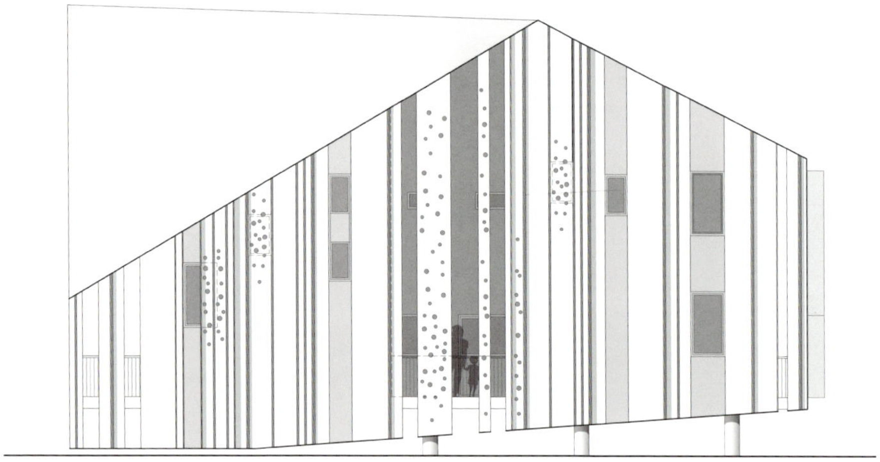

Facade: North/北立面图

Facade: East/东立面图

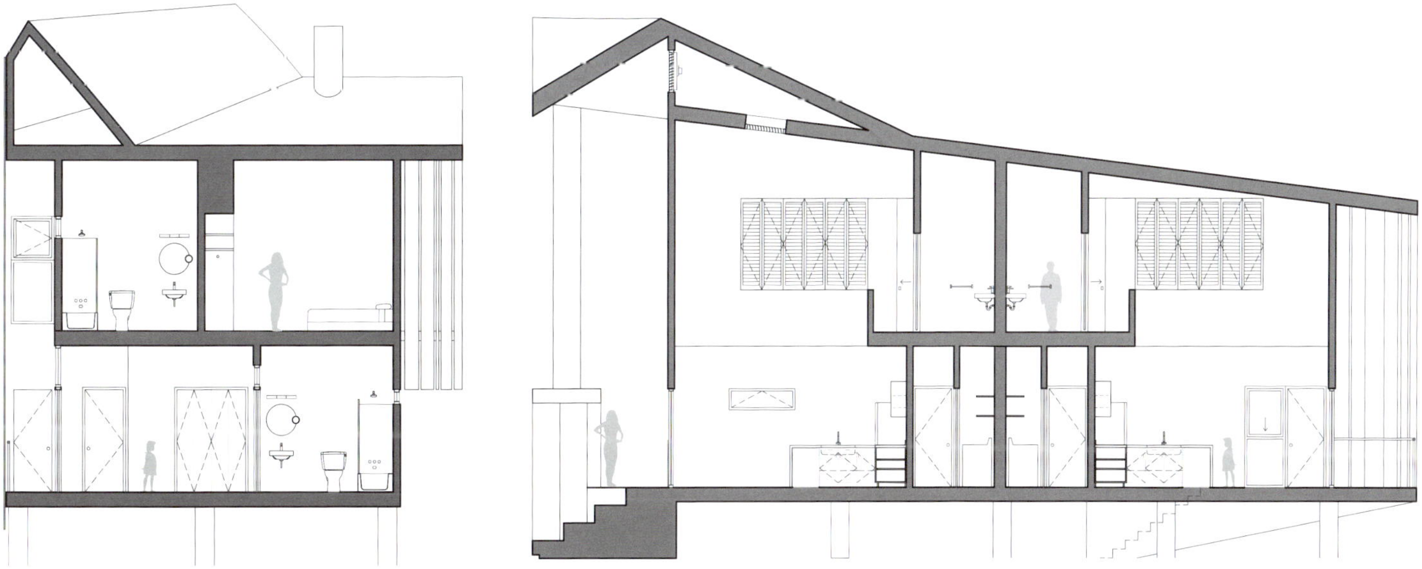

Section/剖面图

Ark House

Axis Mundi

■ Architecture Design/建筑设计: Axis Mundi ■ Project Architect/项目建筑师: John Beckmann, Ronald Dapsis, Masaru Ogasawara, Natacha Mankowski ■ Location/地点: Montana, USA

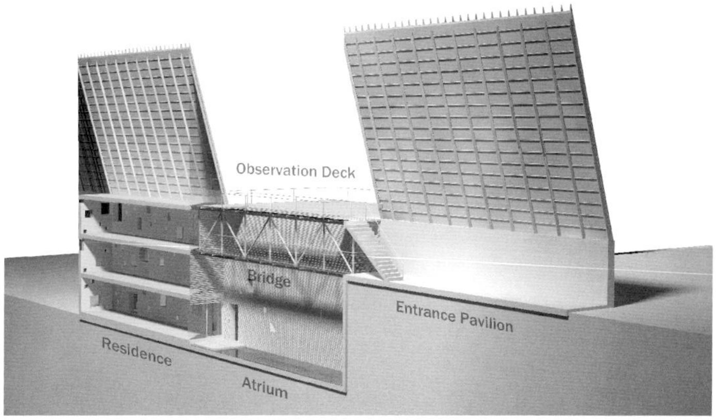

The design for this residence can be likened to the discovery of an archaic sailing vessel, beached on a mountainside, as if a great ocean receded in the ancient past.

Montana's Western mountains have been lifted and folded by plate tectonics and sculpted by glaciers over millions of years. The project is situated on a sloping windswept bluff overlooking Beaver Head National Forest with extraordinary views of Big Sky Mountain.

In a world which is increasingly becoming placeless, our clients requested that we design their home with cultural specificity. It should be of its time, yet be part of the place they love - the vast ancient landscape of Montana.

The overall design is a long barn-like structure bisected across the center by an enormous cantilevered observation deck of nearly 4800 sq. ft.

Half of the main form is an open shell which serves as an entrance pavilion. It contains only a staircase leading up to the observation deck. From the deck, one can enter the main house in the other half. There is an alternate, sheltered path under the deck, opening behind the staircase. Fabricated from Corten steel and glass, a 60-foot bridge spans a 3-story atrium space below.

On the uppermost level of the house is the main living space which includes kitchen, dining and living areas. Two staircases lead down to bedrooms on three lower levels.

All of the floors have breath-taking mountain views at one end and look in to the atrium at the other. The atrium walls are lined with a Corten steel trellis. A lily pond resides at the base.

This home features near-zero energy use thanks to a high performance building envelope, a geothermal heating and cooling system, and photovoltaic panels that produce as much energy as the home needs each year. The 100-acre site preserves the native landscape.

这个住宅设计灵感源于一艘古老的航船——当远古的海水退去后，航船搁浅在了山腰间。

蒙大拿州西部由于几百万年前板块上升和挤压而形成了如今高耸、层叠的山峦地貌。这个项目位于一个荒凉的断崖上，在此处能够眺望远处比弗黑德国家森林公园和大斯凯山的美丽景色。

在世界各地，可供给人类居住使用的土地逐渐减少，由此，项目的委托人要求能够设计出与文化特质相结合的建筑房屋，设计的风格要求符合时代背景，还要成为他们所热爱的古老而又辽阔的蒙大拿土地的一部分。整个建筑的设计是一个长形的谷仓式结构，有一个巨大的悬臂瞭望甲板横跨其中，约有446平方米。

一个开放式的外框占去了该结构的大半部分，里面只建有一个通向观光甲板的楼梯，从甲板的这一侧，可以进入到建筑结构的另一半。在甲板的下面有一个交错的小径，由甲板遮蔽着，出口背向楼梯间。建筑由Corten钢和玻璃建造而成，一个18.3米宽的桥横跨了建筑的三层结构空间。

建筑的最高层，是主要的活动区域，包括了厨房、餐厅和客厅。通过两个楼梯，可以来到下面的卧室。

在建筑中的每一层，一端能看到令人心旷神怡的山峦景观，而另一端则能看到中间的观光甲板。甲板两侧的墙壁是由一条条的Corten钢条组合而成的，在其下端，是一个百合花坛。

这栋房子的特点在于能源的零消耗。这得益于精湛的建筑外形设计，采用地热供暖系统、自然降温系统和光电板发电，这些节能措施维持了这栋房子正常的能耗。40.4公顷的场地为该建筑提供了一个纯自然的环境。

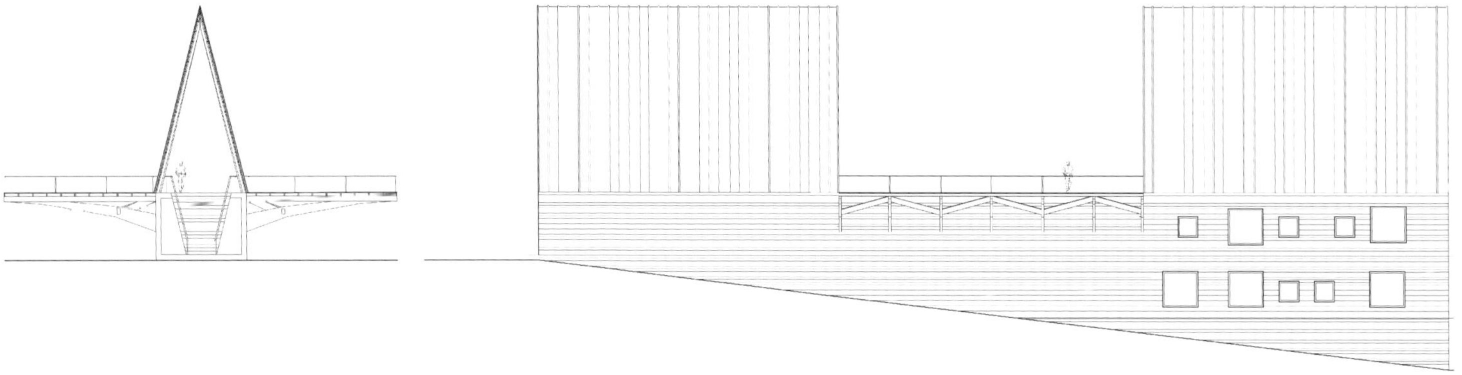

Facade/立面图

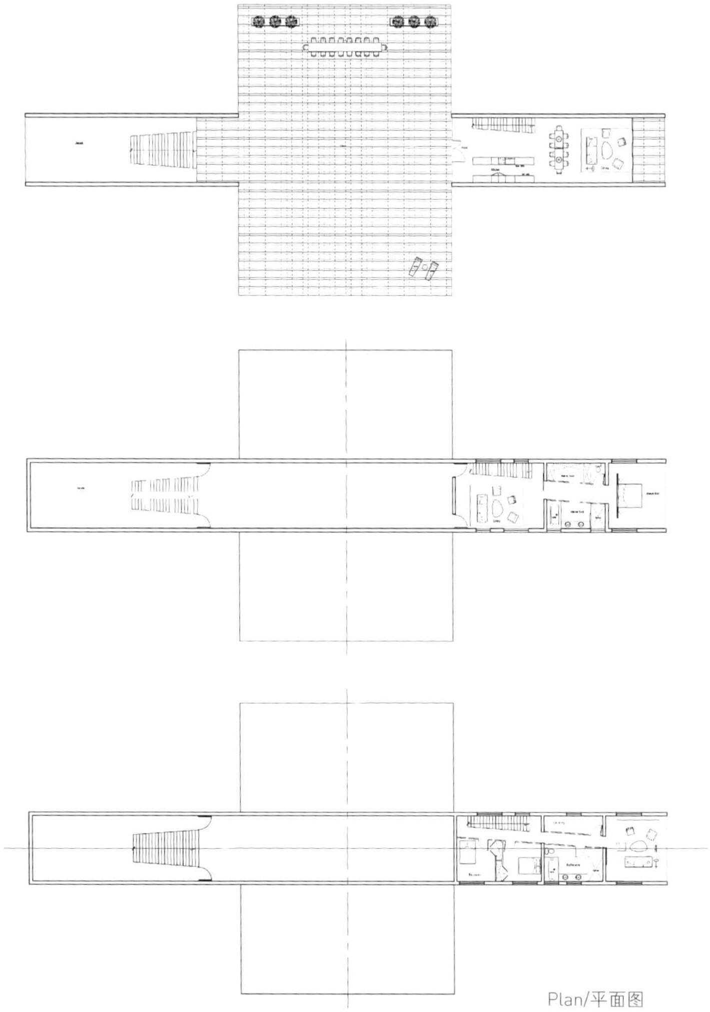

Plan/平面图

Tree House

SHED Architecture & Design

■ Architecture Design/建筑设计: SHED Architecture & Design ■ Project Architect/项目建筑师: Prentis Hale ■ Location/地点: Seattle, USA ■ Area/面积: house 152.7 m^2/carport 37.6 m^2
■ Completion/建成时间: 2009 ■ Photograph/摄影: Steve Hanson

Treehouse is a single family residence located on a park in Seattle, USA. Built on an inaccessible, steep slope lot, the project overcame significant design and construction challenges to create a unique home elevated in the trees.

The project had three primary design objectives: 1) minimize long term site impacts; 2) design a cost effective and resource efficient building; 3) use view, light and enclosure to accentuate the experience of living in the tree canopy.

Site conditions generated design solutions.

- Site accessibility and zoning restrictions necessitated a compact floor plan
- Design solution worked within the limitations of crane to facilitate construction
- Auger cast piles anchor the building to the slope and preserve the natural drainage/slope
- Steel framing support the house and carport above the slope and in the trees

The project pursued sustainable design solutions.

- Compact footprint reduces site impact, as well as material costs / long term energy use
- SIPS roof panels / BIBS insulation used to create a thermally efficient envelope
- Pre-installed Solar Thermal hot water tank, water l ines and LV control wire (Budget prevented immediate installation of this system)
- Future PV system planned for the carport roof (budget prevented installation)
- Operable skylight passively cools/ventilates the house / reduces lighting requirements
- Heat recovery ventilation (HRV) system supplies all mechanical fresh air
- Hydronic radiators used for primary heating system
- Water conserving fixtures and dual flush toilets
- Energy Star appliances
- All paints and finishes low/no voc
- Vintage/salvage materials and furnishings

Treehouse项目位于美国西雅图的一个公园中，是一栋独立的家用住宅，其地势陡峭，不方便通行。该项目的设计克服了重重困难，最终成功地在树林间建造了一个独一无二的房屋。

该项目的设计主要有三大目标：1)尽可能减小对土地的长期影响；2)建造一个价位合理、高效节能的建筑；3)运用视野、光线和周围环境突出在树间居住的感受。

而正是该地的地势条件决定了这样的设计理念。

- 根据地势的可行性和区域的约束性设计一个密集型建筑。
- 设计方案需要考虑建设中施工设备的使用范围。
- 利用钻孔机为房子挖管道，保证利用斜坡自然排水。
- 房子采用钢结构支撑，在斜坡处安置车棚。

该设计追求可持续发展的设计理念。

- 节约地面，减小房屋对地面的影响，减少材料费用和长期的能源消耗。
- SIPS屋顶板面和BIBS保温装置是被用来建造一个高效保温的外形。
- 选用太阳能热水器、太阳能水管和低压控制电源线（没有足够的经费直接满足这些功能）。
- 为车棚顶部计划选择先进的PV系统（没有足够经费）。
- 可操作的天窗能够调节室内温度、保证通风换气、增加室内光线。
- 热回收通风系统可以提供新鲜的空气。
- 液体循环暖气管用做主要的供暖体系。
- 水的双重保护装置和抽水马桶。
- 节能之星的家用电器。
- 所有的油漆选用无挥发性的有机化合物。
- 利用古老的、废旧的材料和家具设施。

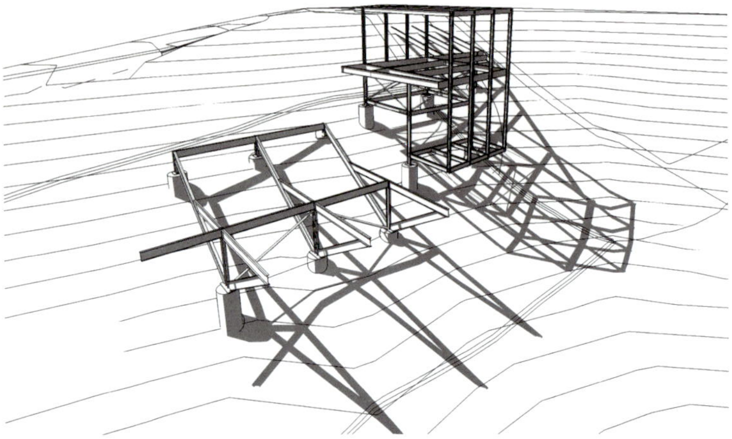

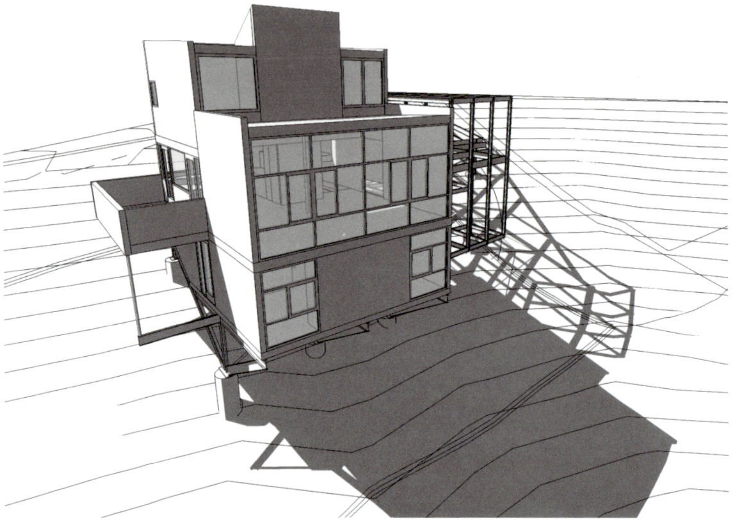

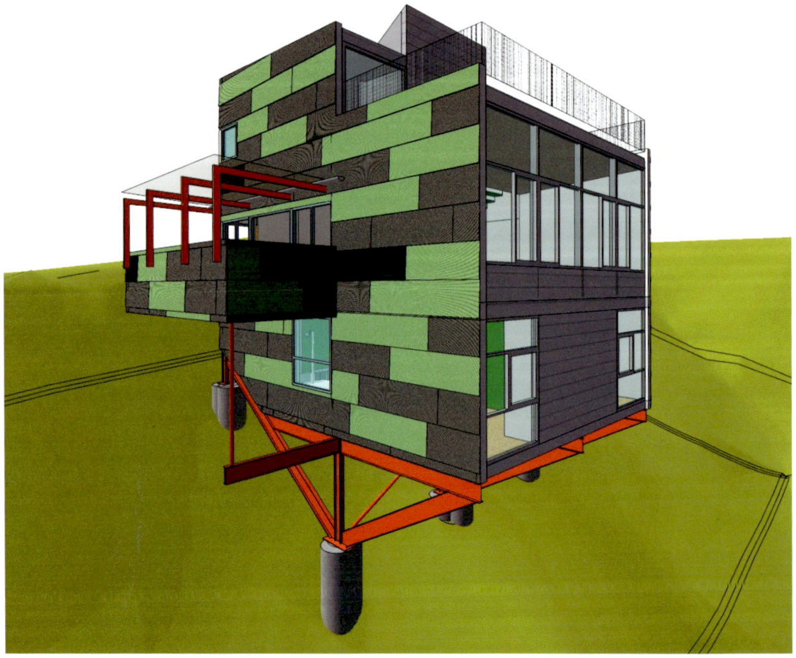

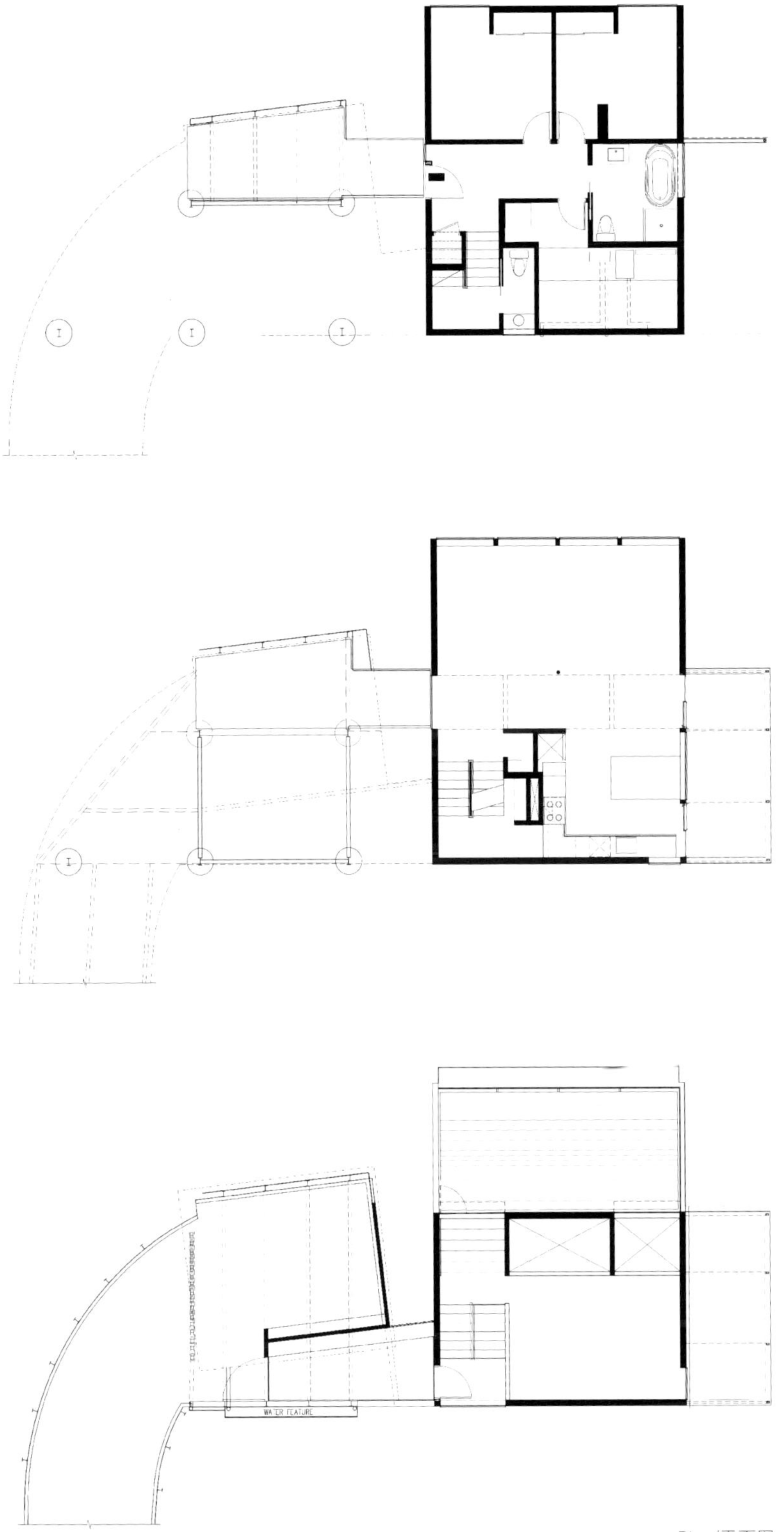

Plan/平面图

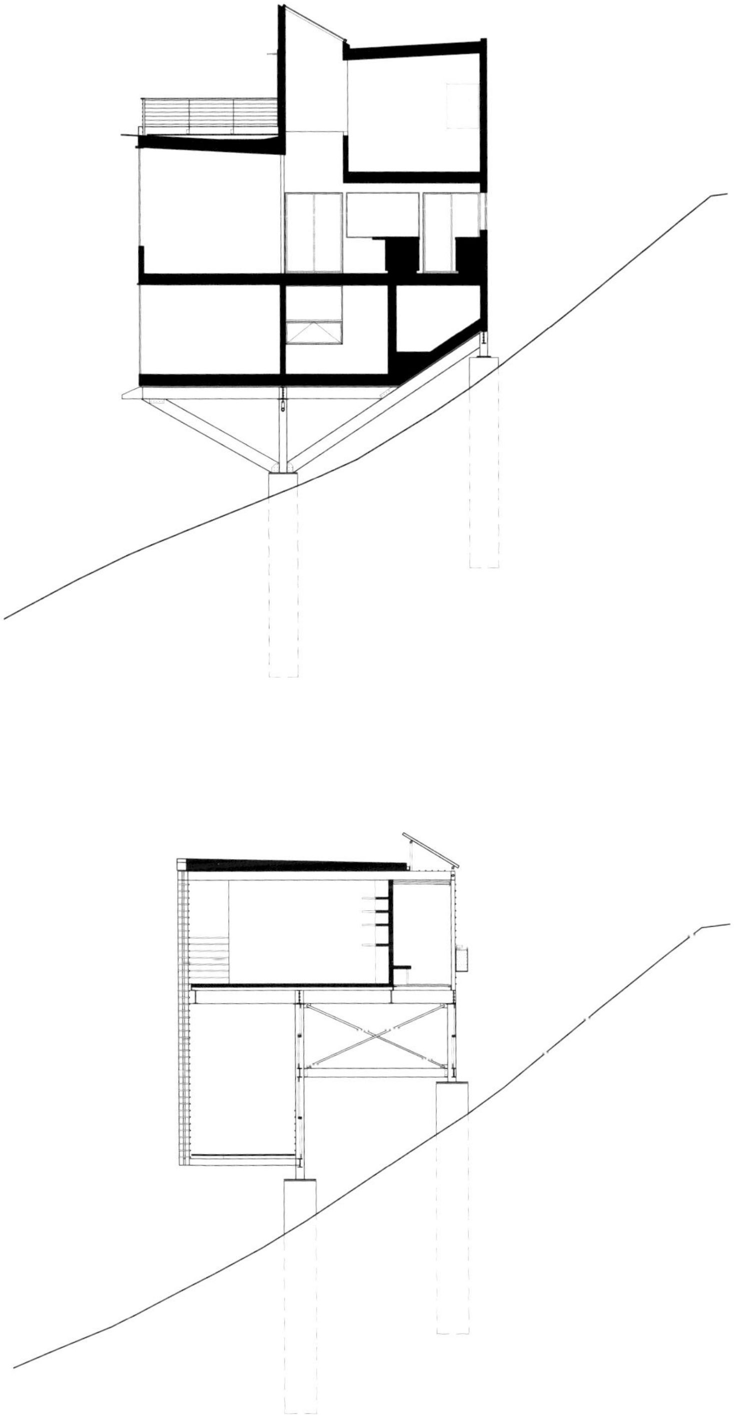

Section/剖面图

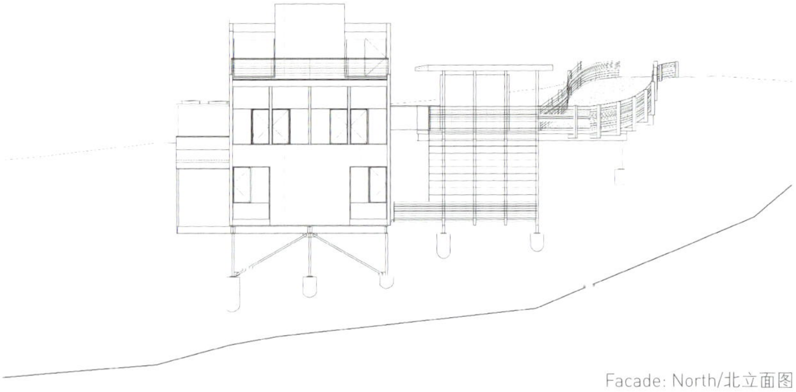

Facade: North/北立面图

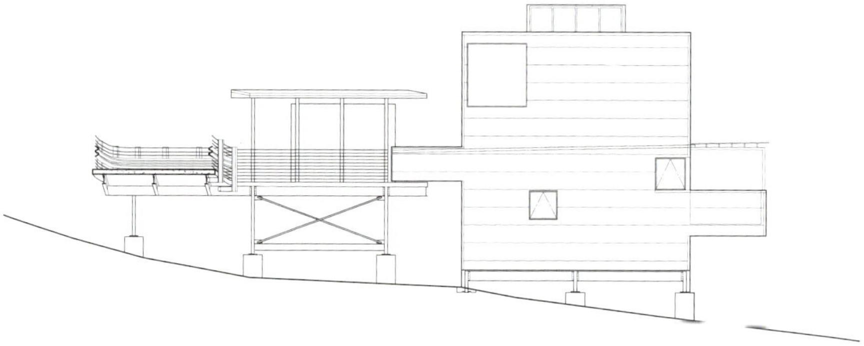

Facade: South/南立面图

Backyard House

SHED Architecture & Design

■ Architecture Design/建筑设计: SHED Architecture & Design ■ Project Architect/项目建筑师: Thomas Schaer ■ Location/地点: Seattle, USA ■ Area/面积: site 276.8 m^2 ■ Completion/建成时间: 2008
■ Photograph/摄影: Benjamin Benshneider

Located in a central but marginal Seattle neighborhood, the Backyard House is a speculative infill development that takes its name from its site—the subdivided backyard of an existing single family house. The projects fundamental objective, to build an affordable house suited to a broad range of potential buyers, resulted in a program that emphasized economical construction, long term operational efficiency, a limited palette of durable materials, and the creation of bright, naturally lit spaces integrated with the site and views.

At the ground floor a series of platforms step down through the interior to define distinct program elements and then re-emerge from the house as a terrace at the edge of the bluff. The intent was to enable inhabitants to experience the topography and for space to flow organically down through the building and its compact site. To define enclosure on these platforms, wall planes and infill glazing are organized to create a sense of both prospect and refuge within the house and site, as well as maximizing access to light in all rooms. Walls, openings and platforms combine to create interconnected spaces that are open and protected, expansive and secluded.

The wood framed structure on concrete slab foundation contains living spaces, a small office, 3 bedrooms, and 2.5 baths on two floors totaling 1960 SF. The stacked stair continues to the roof level where a storage attic and roof terrace is located. The house provides basic amenities such as clear room planning with access to light, ample storage, and spaces that accommodate basic furniture layouts. Program elements were added or combined in small ways to improve use and function, including a large laundry zone in the second bathroom and an entry bench and mail/phone counter at the main entry.

Project detailing and materials were direct, durable and economical. Exterior walls were clad with painted Fiber Cement Panels with areas of wood infill. Aluminum windows were used both for long term durability and to maximize glazing area for interior space. In every aspect the house strived for maximum design impact with a minimum number of materials and custom details.

Backyard House项目位于西雅图中心边缘附近，是一个探索式建筑。它的名字取自它所处的位置——一栋独户住宅划分出来的后院。该项目的基本目标是建造一栋经济型房屋，以迎合众多的潜在购买者。由此，该建筑注重经济实用的建造方式，长期高效地使用，耐久材料以及与该地景观结合的自然采光的应用。

在房屋的一层，一连串的平台一步步向下延伸，旨在使居住者能够体验到该处的地形结构，在房屋之间的区域活动自如。设计平台使用了墙面和内置玻璃作为围栏，从而使住宅有了一定的遮蔽，这样可以保证所有的房间内能够最大限度地采光。墙、入口和平台的巧妙结合，创造出相互连通的空间，既开放又有所遮蔽，既广阔又隐遁。

混凝土板基础上的木质框架结构，包含了起居空间，一个小办公室，三间卧室，两间半浴室，总共两层楼，建筑面积达182.1平方米。楼梯一直持续到顶层，在那里有一个储藏阁楼和屋顶露台。这所住宅提供了简单、舒适的设施，如透光的整理室，宽敞的储藏间和能安放家具的空间。设计增加了一些元素或将其以组合的细节方式呈现，从而增强了效用和功能，其中包括在第二个浴室里的一个大的洗衣区和门口的一个长椅以及主门口的电话柜台。

项目的细节设计和所用材料都是恰到好处，耐久并经济。外墙覆盖着纤维水泥和木质填充物，而使用的铝合金窗户，是出于经久耐用和将室内空间玻璃面积最大化的考虑。设计师在各个方面都力求定制细节、物美价廉，为该项目做出最好的设计。

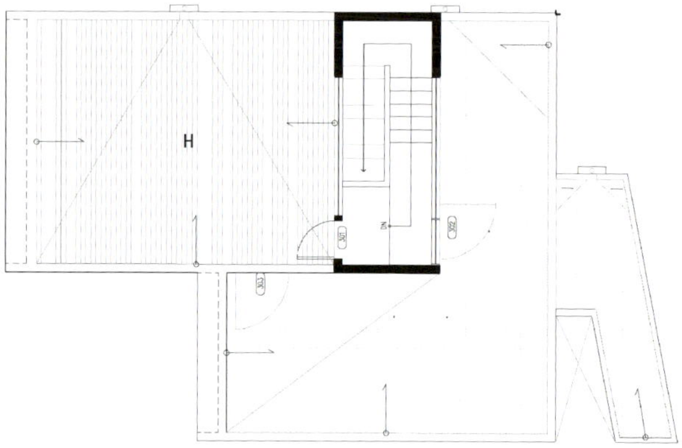

Roof Floor Plan/屋顶平面图

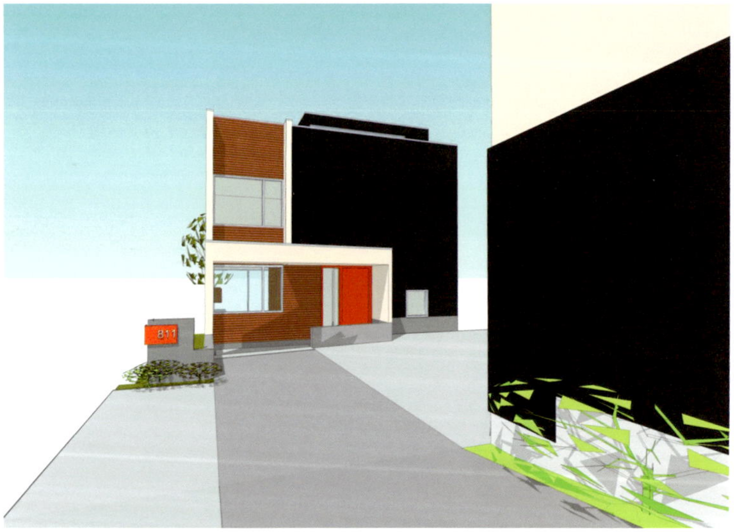

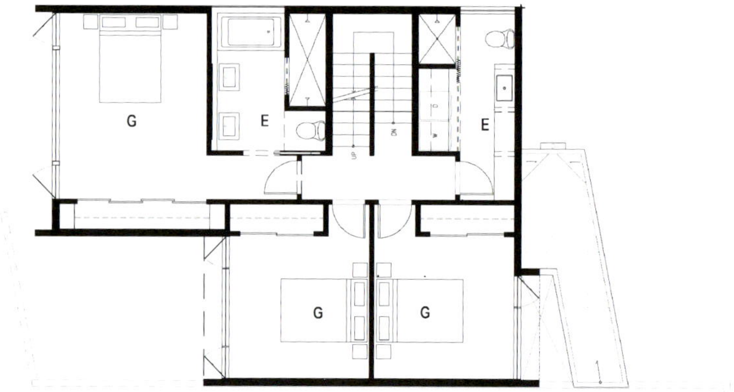

Second Floor Plan/二层平面图

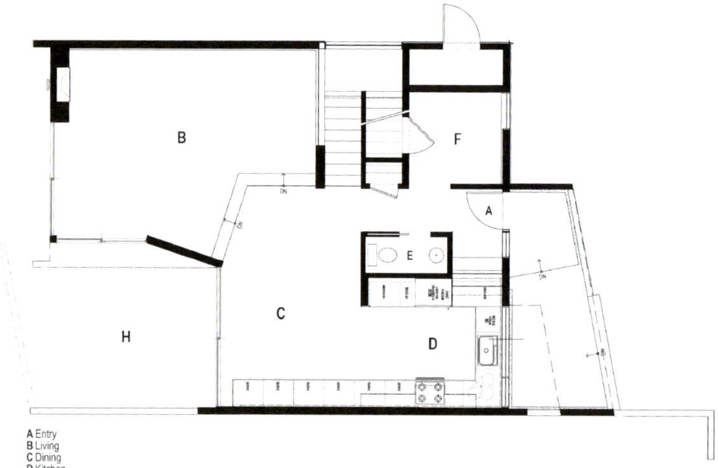

A Entry
B Living
C Dining
D Kitchen
E Bathroom
F Office
G Bedroom
H Patio / Deck

First Floor Plan/一层平面图

Facade: East/东立面图

Facade: West/西立面图

Facade: North/北立面图

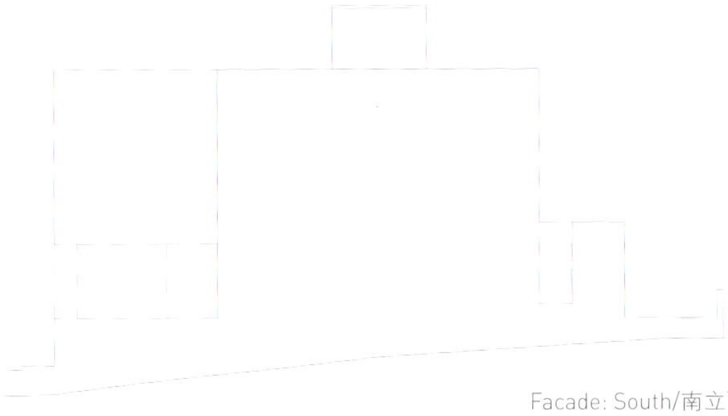
Facade: South/南立面图

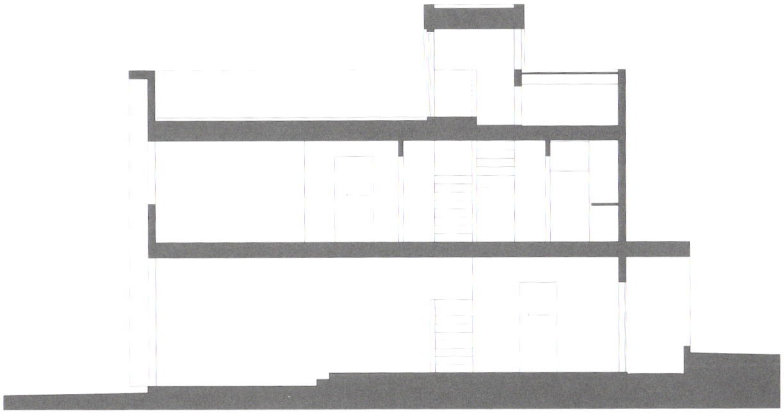
Section/剖面图

Anglesea House

Andrew Maynard Architects

■ Architecture Design/建筑设计: Andrew Maynard Architects ■ Project Architect/项目建筑师: Andrew Maynard, Mark Austin, Matt McClurg ■ Location/地点: Victoria, Australia ■ Area/面积: 150 m^2
■ Completion/建成时间: 2009 ■ Photograph/摄影: Peter Bennetts

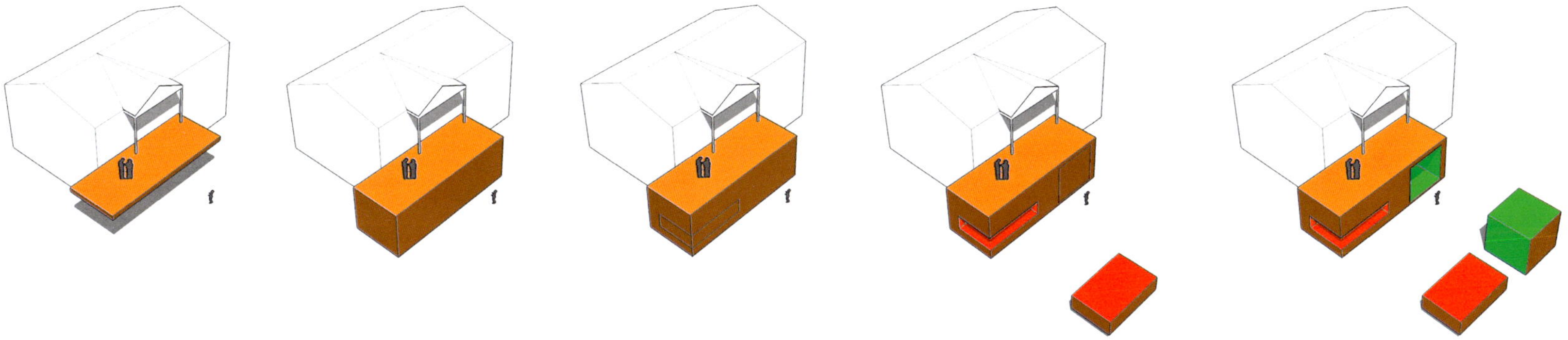

More frequently holiday homes are becoming little more than transplanted suburban ugliness; the great Australian tradition of the 'shack' is in danger of being superseded by bloated mansions with four bathrooms and all the trappings of modern life. With this project we set out to celebrate the holiday shack without adopting shack typology and as such we have kept close to the original building's footprint to avoid taking over the rugged coastal block.

Our clients approached us with a brief for 'much more room' for their 'aging' family holiday home. The design response was a series of finely-crafted timber boxes nestled around the existing house.

The Northern addition replaces an old timber deck that previously divided the two storeys and radically reduced sunlight to the living area on the lower level, making the space beneath damp, dark and disconnected from the rest of the house. The trafficable roof of this addition is now extruded down to the earth, creating a 3 metre thick deck and grounding the entire house to the site, while extending the top floor living spaces out into the treetops.

This "thick deck" was then carved out to allow the space within to become habitable play spaces. What could have been a simple suspended deck structure now becomes a multi-dimensioned element, also serving as a roof for the new living/space area below.

The spaces within this timber box to the North can be opened up entirely to the surrounding bush block. Sliding windows above the daybed are concealed within the structure when open & full height glazed doors to the North & West allow sun to penetrate deep into the interior. These extensive openings allow maximum physical and visual interaction between this internal environment and the surrounding bush land, whether views externally from within the space or from the block looking back into the boldly colour filled internal spaces.

越发增多的度假住宅，日益成为郊区中一幕幕移植的景观，显得那么不伦不类。澳大利亚一贯的“棚屋”传统日渐消隐，已被那些带有四个卫生间的奢华豪宅以及现代生活的繁冗细节替代。而这个设计，使我们眼前一亮，仿佛看到了度假棚屋的回归，虽然设计没有采用棚屋的类型，但很大程度上延续了传统建筑风格，尽量避免了人类对这片原生态的海岸地带的侵犯。

会面时，客户向我们提出了一个简明的要求：为他们“年老”的家庭度假住宅设计出“更多的空间”。于是，我们设计出了一系列工艺上乘的木箱，安放在现有房屋附近。

房子北面设计加建的结构，替代了那个原本的木质天台。从前，天台将房子隔成了两层，严重地阻挡了底层起居区域的光照，导致底层阴冷潮湿，也与房子的其他空间隔断了。而设计加建结构的屋顶，可直通向地面，这样创造出了一个3米厚的平台，在保证该建筑与地面相连通的同时，将上层的起居活动区域扩展到了树顶的高度。

接下来，设计师对这个“厚重的平台”进行发掘，使其内部空间变为可供居住的娱乐区域。一个原本简陋的悬空平台结构，如今已变成一个有着多重空间的重要元素，同时也充当了较低层起居活动空间的屋顶。

朝北的木箱内部，可以面向周围的灌木林完全敞开。当玻璃门完全敞开时，沙发床上方的滑动窗户就会隐藏起来，这样，阳光就照进房间。不管是从室内往外面看，还是从灌木林向这个色调鲜明的室内望去， 室内环境都与周围的灌木林之间和谐共生，交相辉映。

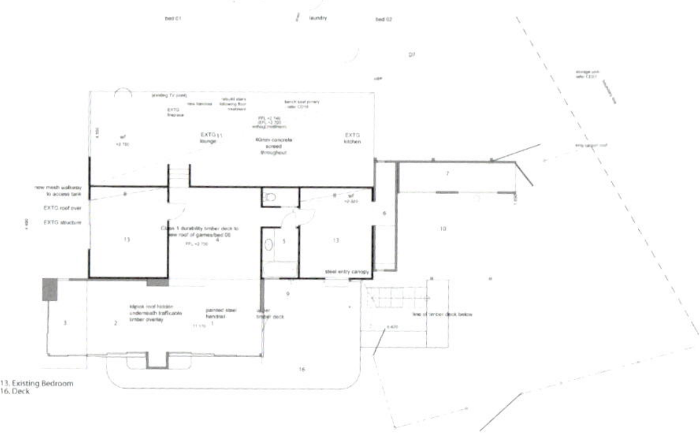

1. Games room
2. Bedroom
3. Daybed
4. Family Room
5. Bathroom
6. Bunk Box
7. Storage
8. Robe
9. Surf Shower
10. Existing car port
11. Subfloor of existing building
13. Existing Bedroom
16. Deck

Second Floor Plan/二层平面图

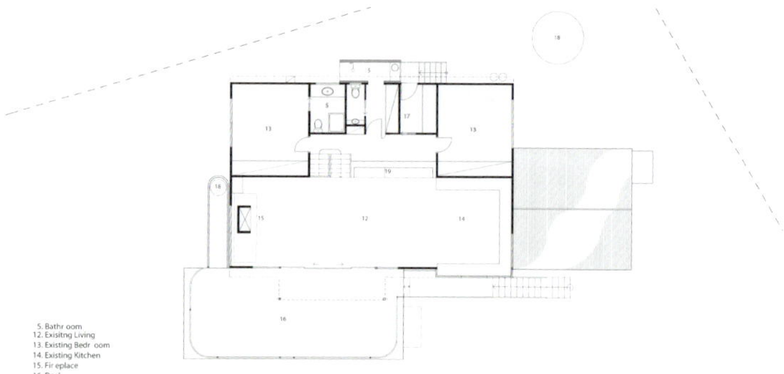

5. Bathr oom
12. Exisitng Living
13. Existing Bedr oom
14. Existing Kitchen
15. Fir eplace
16. Deck
17. Laundry
18. W ater tank
19. Joinery

First Floor Plan/一层平面图

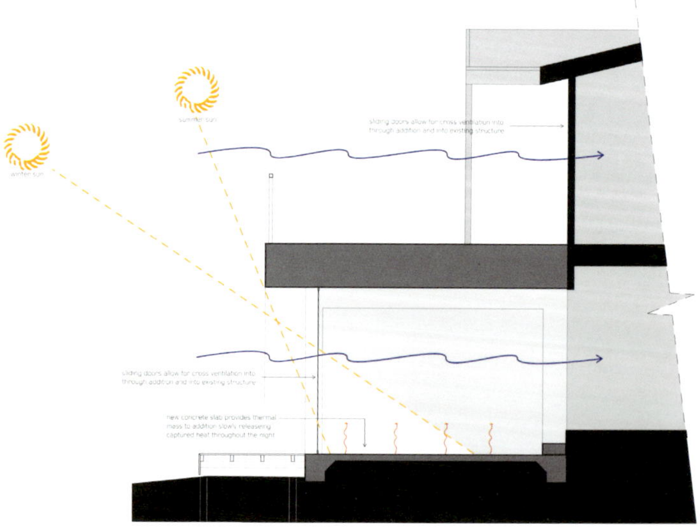

Section/剖面图

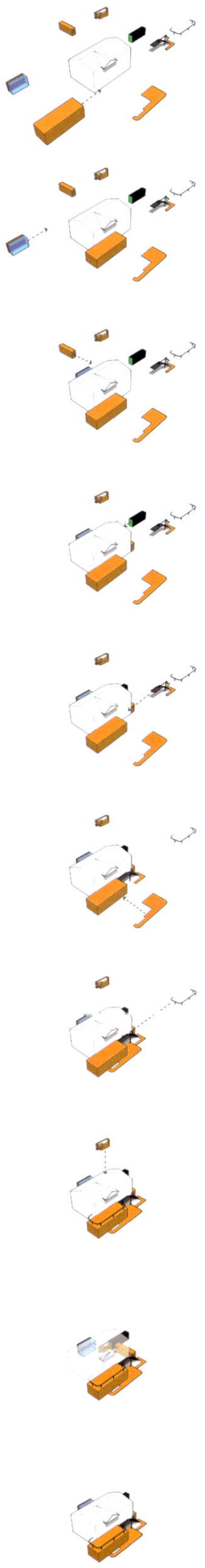

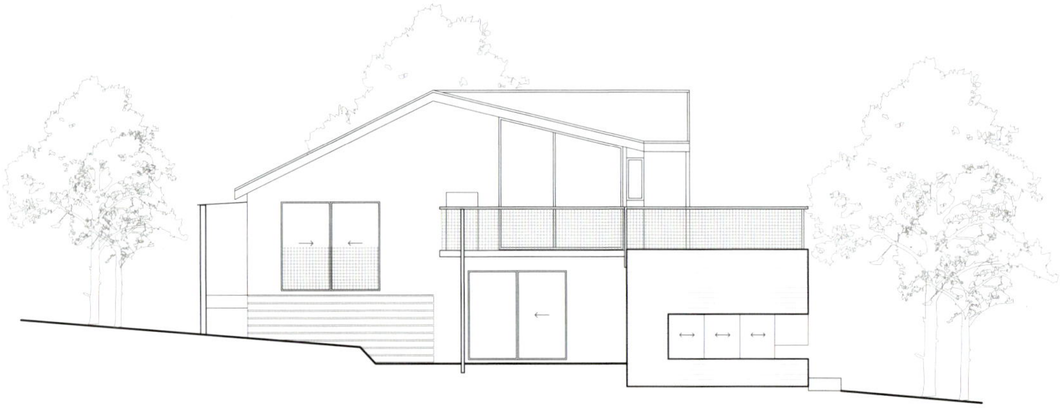

Facade: East/东立面图

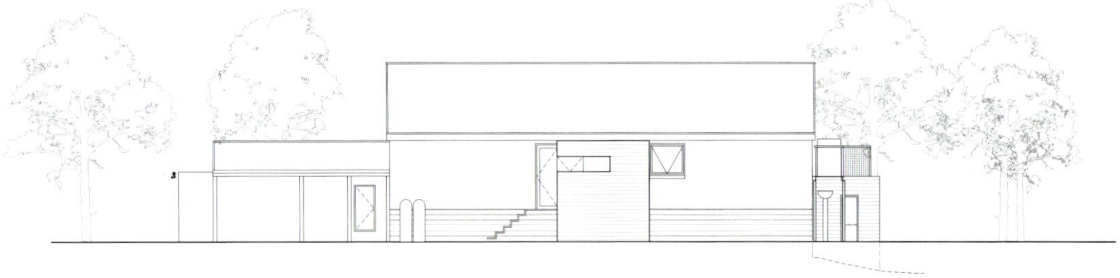

Facade: South/南立面图

Facade: West/西立面图

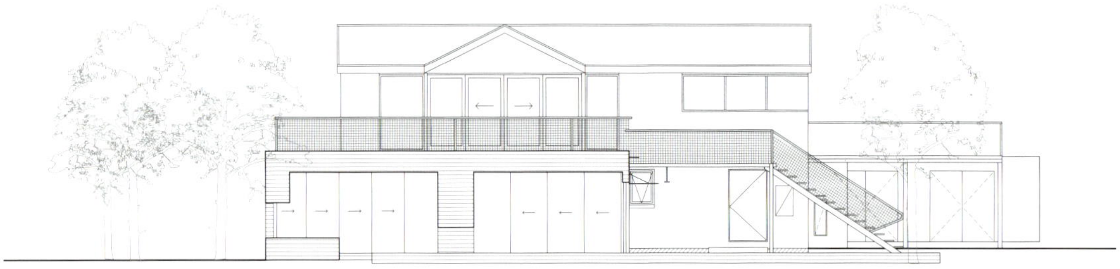

Facade: North/北立面图

Barrow House

Andrew Maynard Architects

- Architecture Design/建筑设计: Andrew Maynard Architects
- Project Architect/项目建筑师: Mark Austin
- Location/地点: Victoria, Australia
- Area/面积: 235 m^2
- Completion/建成时间: 2008
- Photograph/摄影: Peter Bennetts

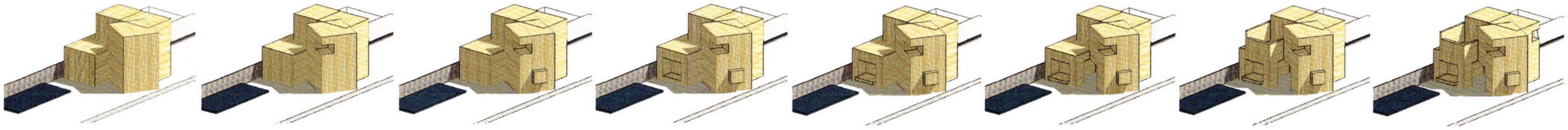

The Barrow extension appears as an arrangement of timber boxes, each independently rotated and subjected to varying amounts of extruding and manipulating forces.
Frequently the windows are setback within the frame of the wall, sometimes flush and occasionally extruding beyond the timber frame. This unconventional approach to massing and window design subverts the conservative planar nature of a 'box'.
The old living space now is occupied as a bathroom. Here, there is no evidence of conventional bathroom materiality, the room still maintaining its appearance as a living/dining space. The only defining feature a free standing cast iron bath (reused from the original demolished bathroom) at the centre of the room, demonstrating the capability of a single element to alter the program of a space. The strategic placement of a separate living space at the western end of the site reflects the focus of the site internally, frames the large open area & increases privacy levels.
Definition between indoor and outdoor is blurred by the transparent divisions of bifold doors and large windows; visual interaction is constant. This central outdoor space becomes part of the living circulation space as the diurnal patterns of the occupants see them traverse the yard to the rear living quarters.

Barrow项目加建的建筑物看起来就是一个木箱子的组合，而每一个木箱都独立地扭曲着，同时承受着多种挤压。
该设计是对传统的木质结构特性的极大挑战。不同窗户的设置以及窗框安装的工艺，进 步提刀了环境的动态感以及变化性。通常情况下，窗户会安放在墙体结构缩进一些的位置上，有时会和墙体齐平，偶尔会超出木框之外。然而这个反常规的做法和窗户的设计却颠覆了“箱子”的平面特性。室外的木材铺贴包裹了房屋，使该建筑物与原房屋融合起来。原先的起居区现在也变成了卫生间，在这个设计中，看不到任何常规的卫生间的痕迹，因为它还是保持着起居室或餐厅的原有风貌，唯一具有辨识度的是一个放在房间中央的落地式的铁质浴盆（由原有废弃的卫生间重新利用），这种设计展示了设计师利用单一元素改造空间的能力。
在项目的西侧巧妙地安放了一个独立的起居区域，反映出室内空间的中心所在，设计师构建了大片的开放区域，同时提升了起居区域隐秘性。
透明的分格折叠门和巨大的窗户，模糊了室内、室外的界限，给人们带来全新的视觉感受。而位于中央的户外活动区域则被设计在主人的日常起居活动范围内，这样主人可以从后院径直来到后面的生活区。

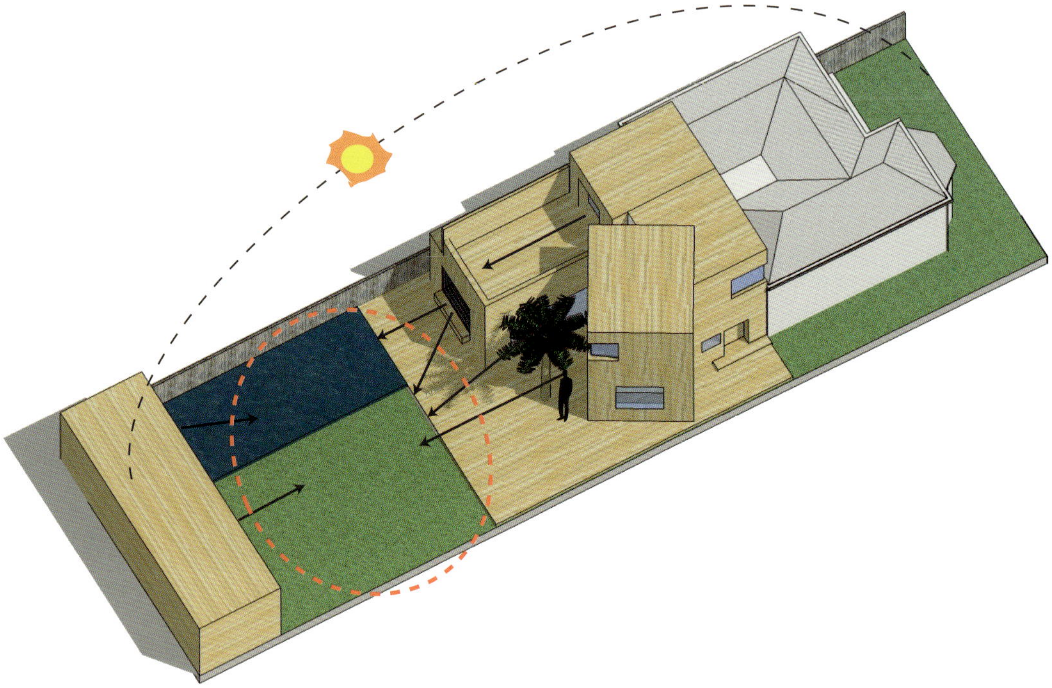

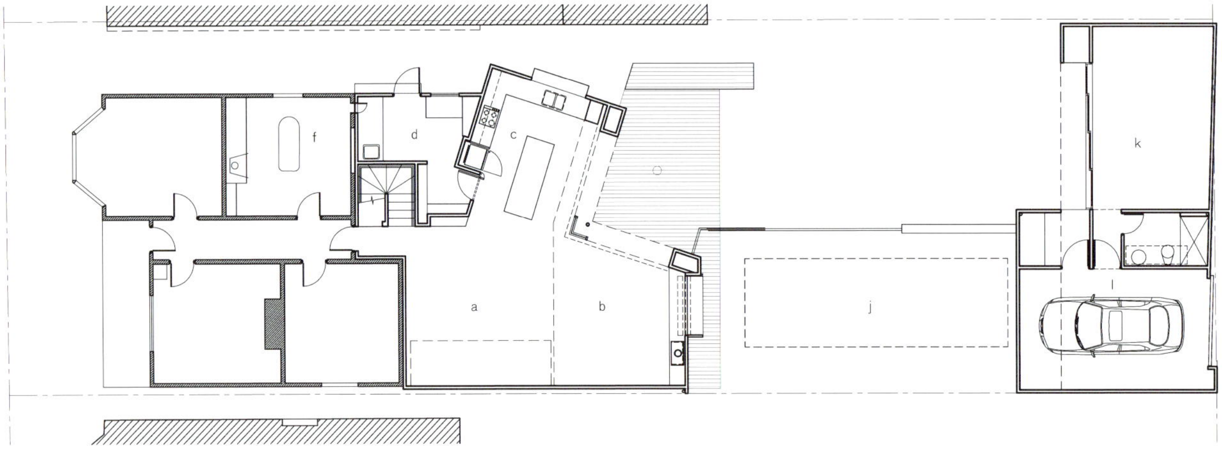

a. living
b. dining
c. kitchen
d. laundry/pantry
e. landing/hallway
f. ensuite/bathroom
g. bedroom
h. deck
i. terrace
j. pool
k. studio
l. garage

First Floor Plan/一层平面图

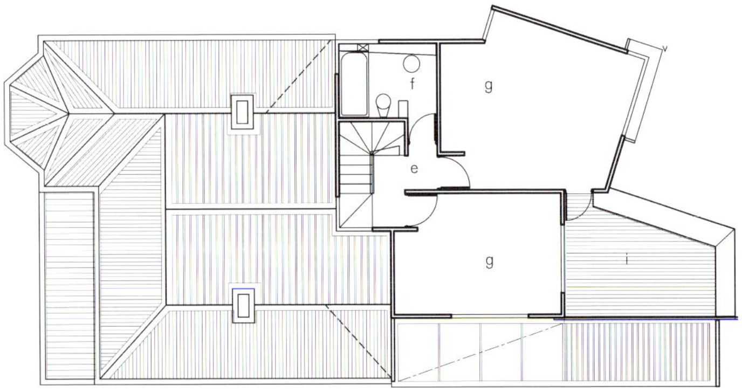

a. living
b. dining
c. kitchen
d. laundry/pantry
e. landing/hallway
f. ensuite
g. bedroom
h. deck
i. terrace
j. pool
k. studio
l. garage

Second Floor Plan/二层平面图

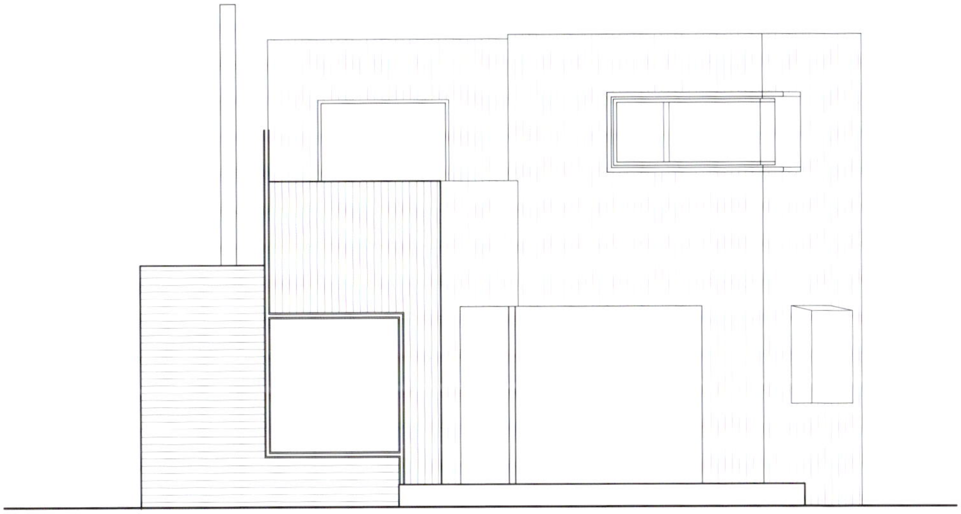

Section/立面图

Tattoo House

Andrew Maynard Architects

■ Architecture Design/建筑设计: Andrew Maynard Architects ■ Project Architect/项目建筑师: Andrew Maynard, Brad Holt, Matt McClurg, Cara Wiseman, Mark Austin ■ Location/地点: Victoria, Australia
■ Area/面积: 72 m^2 ■ Completion/建成时间: 2007 ■ Photograph/摄影: Peter Bennetts

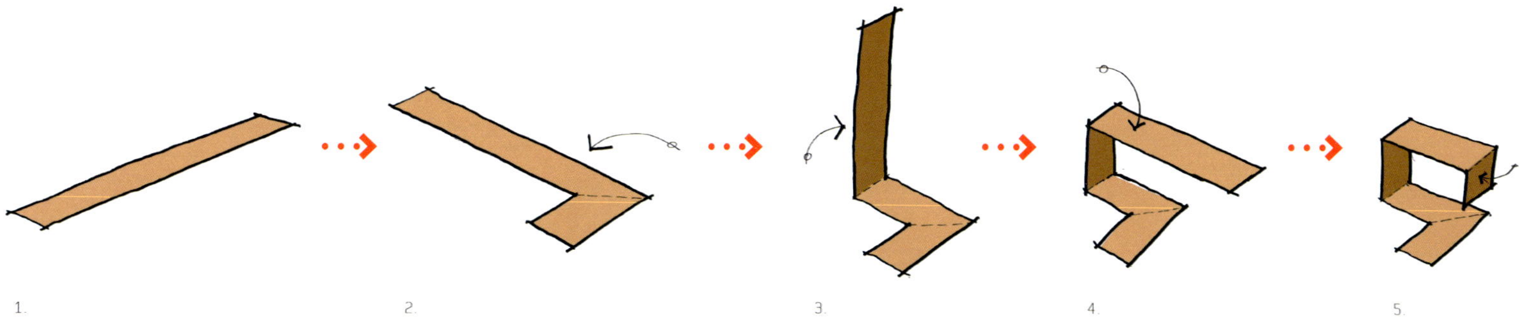

The Tattoo house is Andrew Maynard's latest built project. The building is a small extension to an existing 3 bedroom house in Fitzroy North, Victoria.

The client's brief was delightfully loose; provide new living and kitchen space for a growing young family and create an open plan with plenty of natural light and high ceilings. A kitchen and a flexible work space incorporated into the extension rounded out the brief. By balancing these desires against the various restrictions of budget, town planning requirements and available space, tactics were employed to make everything within the space perform multiple tasks, both functionally and conceptually.

The tight budget generated many of the design decisions. The form is a simple box- the strongest form an architect can achieve at a bargain basement price. From the starting point of the simple box we began to express the addition as a covered external space. The original house is very internalised. The extension was to be as open as possible so that when one crosses the threshold between the original and new structures they felt as though they were entering a covered deck rather than additional internalised space.

Generally we try to avoid having separate ideas performing different functions. We try to be economic with our concepts and make a couple of key ideas achieve multiple and varied tasks. Every element within the Tattoo House needed to perform multiple functions for maximum return- hence the kitchen bench becomes part of the stair, and the screening required by council reflects heat and glare away from the expansive windows. Despite these challenges a double story, non-domestic scale space was achieved.

这栋有文身的房子是Andrew Maynard建筑设计事务所的最新力作。该建筑位于维多利亚的菲茨罗伊北部，是一栋已有3个卧室的住宅的扩建项目。

客户的要求倒是很宽泛，要求为一个年轻的家庭设计出全新的起居和厨房区域，并营造出一个采光效果极佳的高屋顶效果。后来又补充说明，厨房和一个灵活的工作区也要纳入该扩展项目中。

在预算、城市规划要求以及可利用的空间等限制条件下，设计师们巧妙地在既定空间范围内，使各个设计单项不管是在功能上还是概念上都发挥了多重功效，最大程度上满足客户的需要。

紧缩的预算成本决定了该设计的多项决策。房屋的结构酷似一个简易的箱子，这是设计师在仅够建造一个地下室的预算的情况下，所能设计出的最稳固的结构。从这个简易的箱子开始，设计师就力求将该加建建筑物设计成一个有屋顶的开放式空间。原来的房子很封闭，而加建的建筑物就要尽量开放，这样，当穿过原有建筑和新建筑之间的走廊的时候，就会让人感觉到他们好像是在进入一个有屋顶的平台，而不是另一个封闭的空间。

通常情况下，设计师会尽量避免用多个孤立的想法来分别实现多种功能，而是采用最经济实用的想法，通过一系列重要的构想，完成多种不同的任务。这栋有文身的房屋里的每个元素，都需要相应地发挥多重功效，由此，厨房的长椅也变成了楼梯的一部分，设计顾问组提出要采用的遮蔽物，也可以反射热量并且成为大窗户之外的另一道引人注目的风景。尽管面对如此挑战，设计师还是把一个双层非居住规模的空间成功地修建出来。

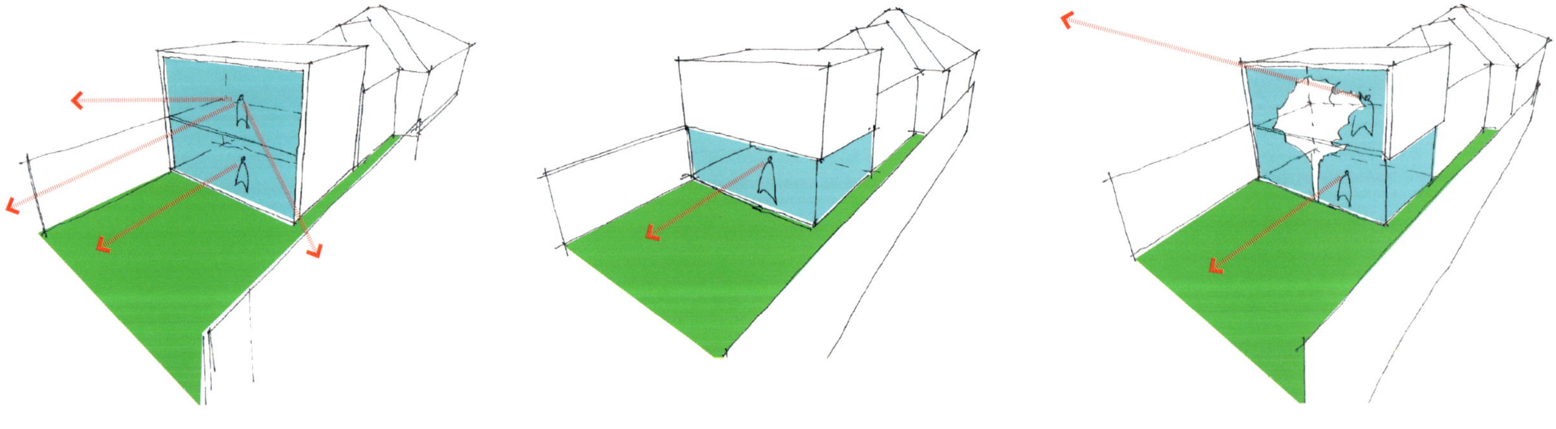

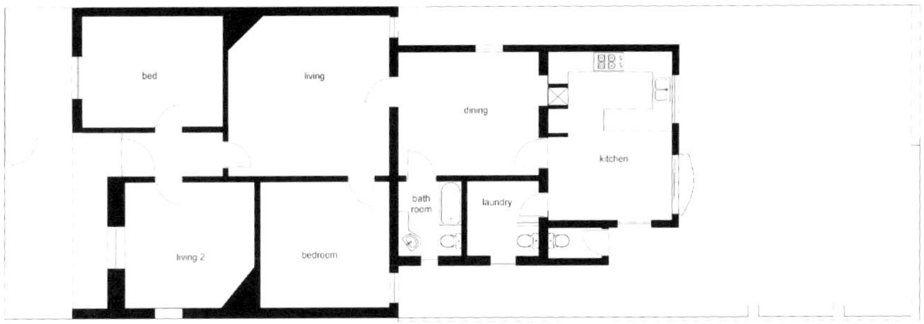

Origina Floor Plan/原平面图

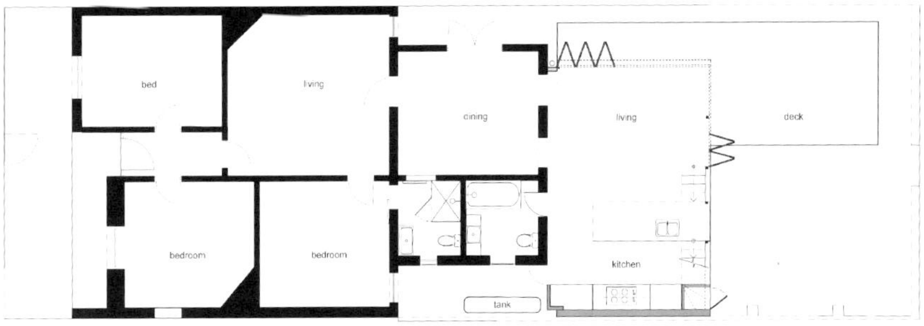

Ground Floor Plan/底层平面图

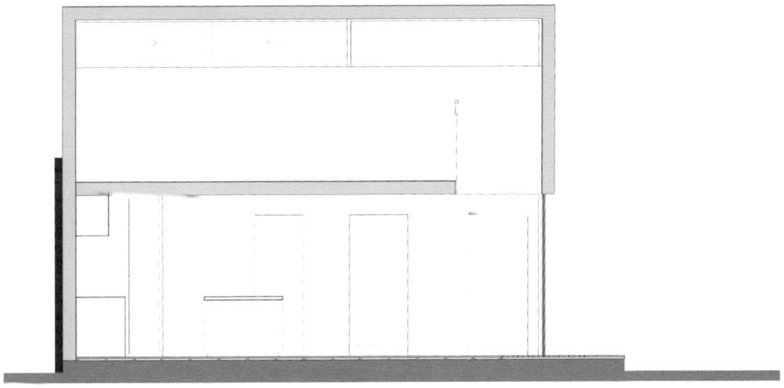

Section/剖面图

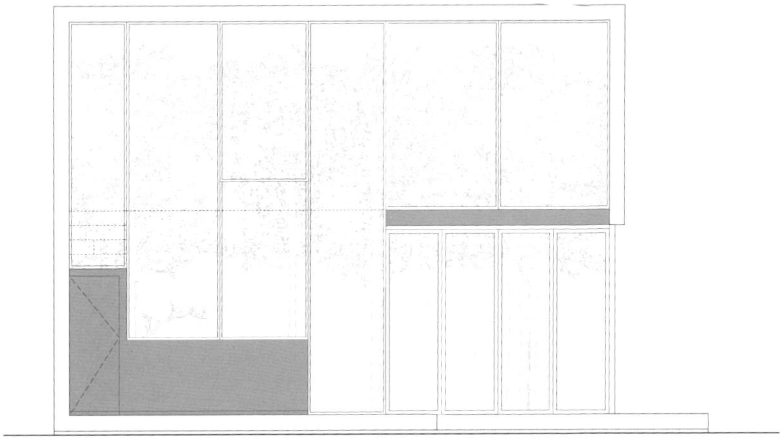

Facade/立面图

Y House

asensio_mah

■ Architecture Design/建筑设计: asensio_mah ■ Design Team/设计团队: Leyre Asensio-Villoria, David Mah, Taewoo Kang, Juhyun Kim, Roberto Leiro Ozores, Andres Mendoza ■ Location/地点: Perth, Australia
■ Area/面积: 285 m²

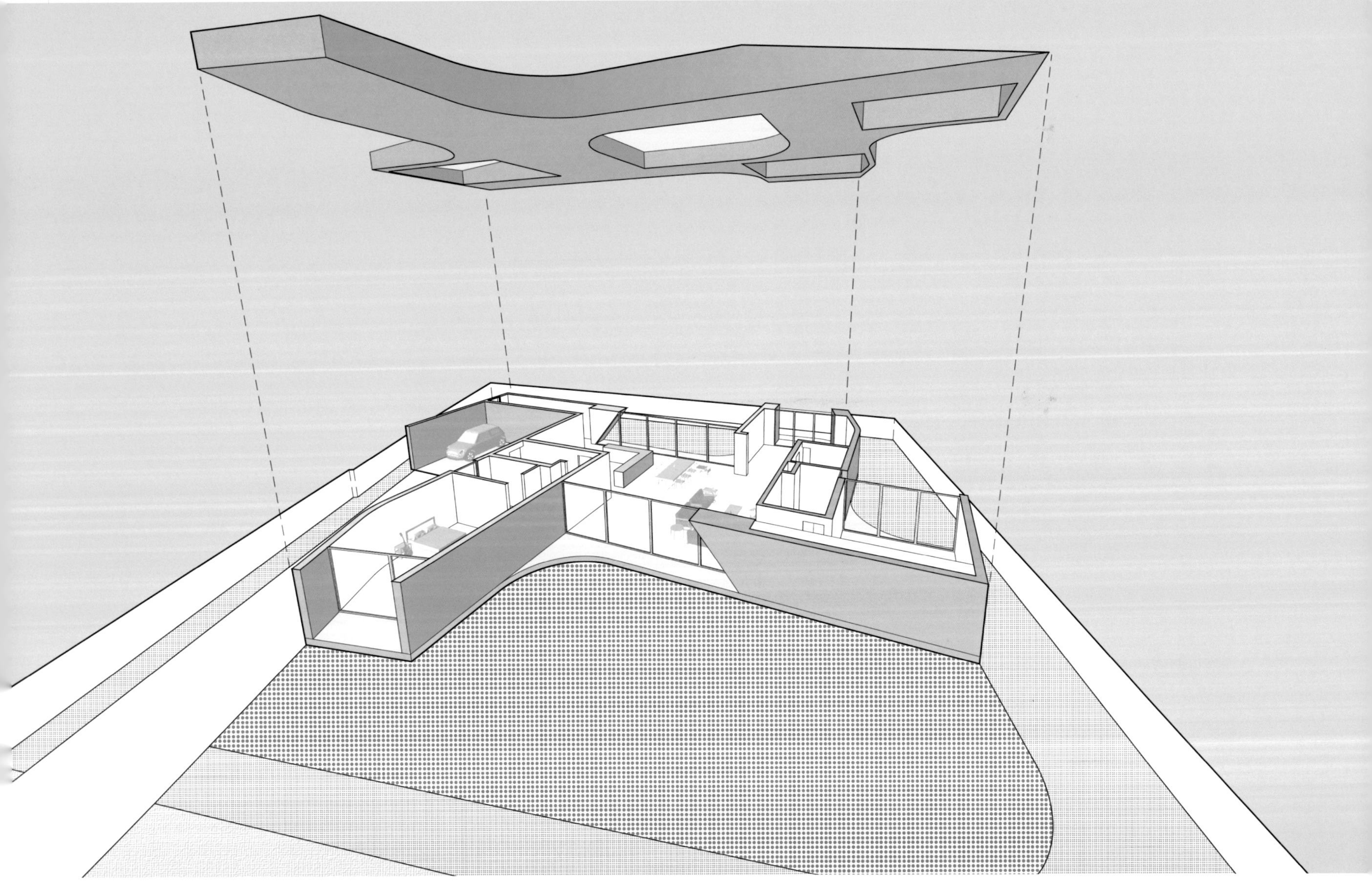

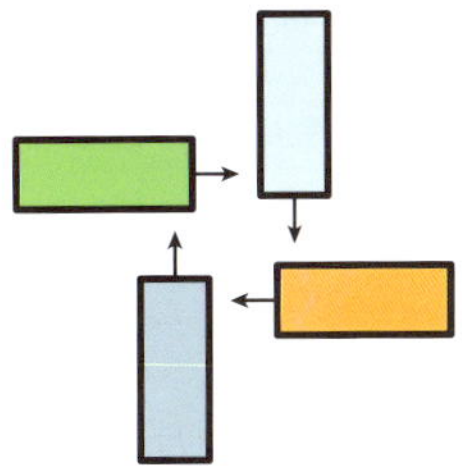

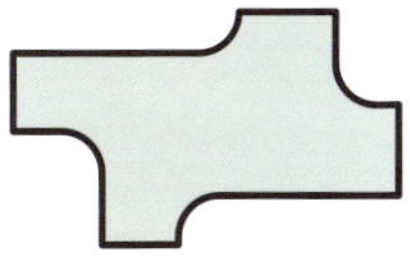

Y House is for a family with a specific lifestyle desire, which is in many ways a universal problem in the definition of domestic environments. The couple and their son live in the house as a family unit but are in all respects fiercely independent individuals. Their simultaneous desire for a shared living experience as a family with an equally strong desire for a high degree of independence and autonomy as individuals was a tension within the brief that needed to be addressed in the architectural and landscape organization of the domestic environment.

Typically, houses within these Australian suburbs, are by default maximized to the full capacity allowed by planning guidelines and regulations. These almost always result in oversized houses that leave a minimal outdoor area for gardens and are often simply the result of setback regulations. In Y-House, the design of the house needed to take the definition of its outdoor areas into equal consideration. To balance the need for a celebrated communal living area and the autonomy of the private areas of the couple, their son and visiting guests, a pinwheel plan was adopted as the basic diagram for the house organization where each wing of the pinwheel corresponds to the private areas of the house as well as one of the wings dedicated to the service areas o the kitchen, garage, cellar and storage.

The central area, around which the four wings are distributed, is a large living and dining space which is able to extend in the instances where the guest wing is unoccupied. Definition of the outdoor areas into different zones that are clearly framed and segmented by the wings into various private gardens as well as larger front and back gardens that form a continuity through the central living space are a result of the adoption, adaptation and transformation of the pinwheel plan type.

Y House项目的设计，旨在解决客户家庭特殊生活方式的问题，这也是在所有的家庭生活中普遍存在的问题。这对夫妻和他们的儿子虽说是一家人，但在生活方式上却是非常独立的。即使他们居住在一起，也同样保持着高度的独立和自主意识，因此客户要求在建筑设计和景观布局方面体现出他们的理念。

一般来说，在澳大利亚的郊外，建筑设计师会根据规划指导和细则最大化地利用空间。这样一来，设计师经常会设计出过大的房屋，导致户外花园的空间被压缩到最小。Y House项目的设计则将室内空间与室外布局等量分配。为了满足公共区域和夫妻私人空间的不同需求，设计师将建筑的布局设计成风车的形状，这样他们的儿子和来访客人的房间都可以分布在风车翼的位置上，同时将厨房、车库和储藏室等设施设置在内。

由四扇风车翼包围着的中央区域，成为室内开阔的客厅和餐厅,一直延伸到客房之外。风车翼清晰地构建和分隔了多个不同区域，由此，每一个风车翼的外面都有一处私人的小花园，而且房子的前后各有一个花园，这样形成了一个以中央客厅为中心的一体式格局，充分体现了设计师将风车式结构进行成功转变的设计能力。

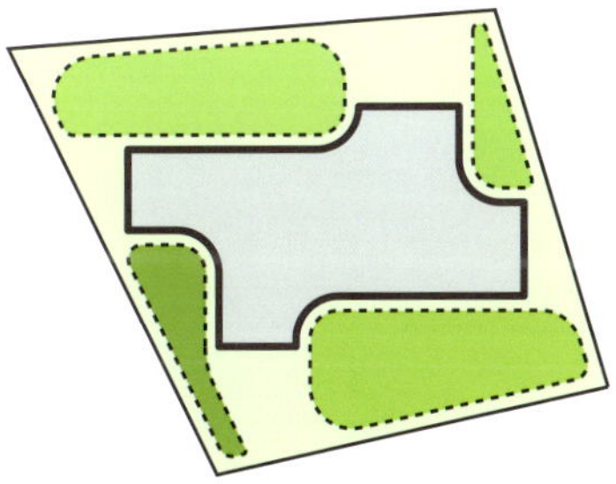

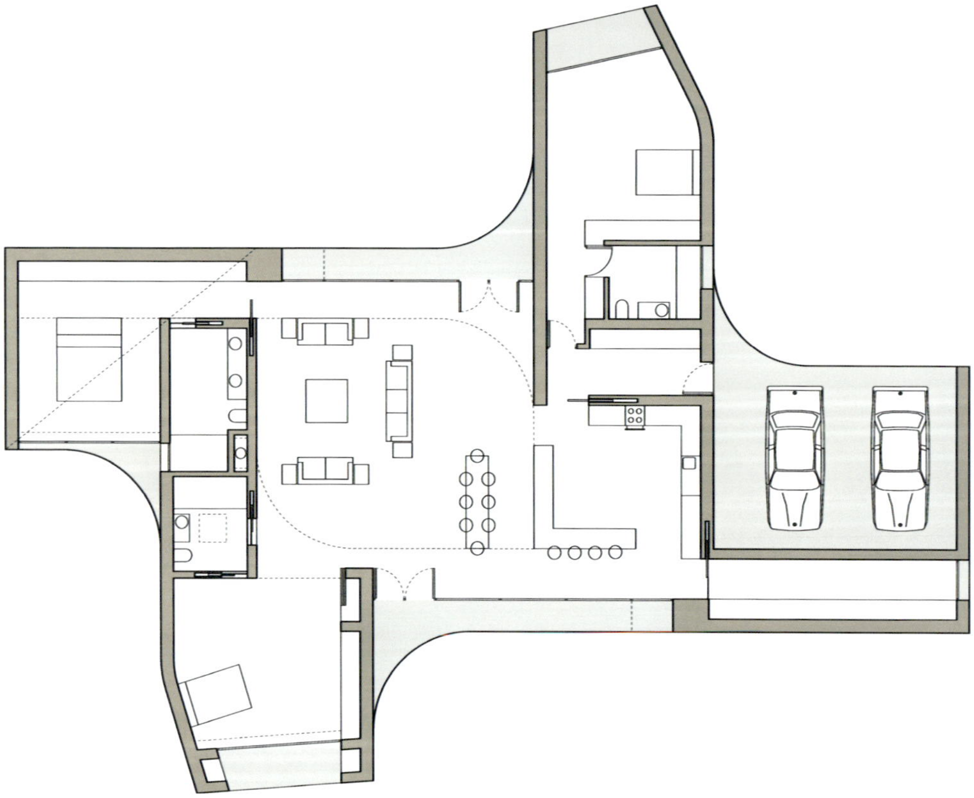

Plan/平面图

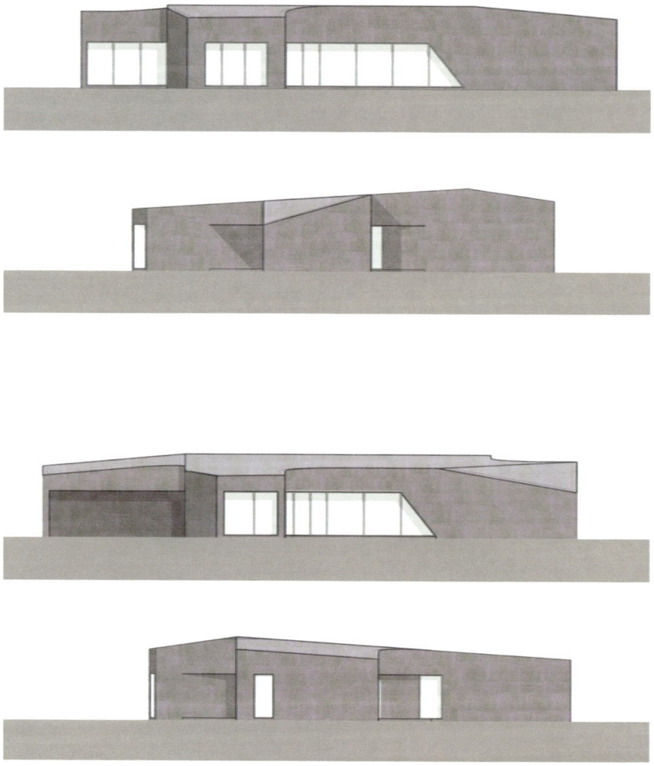

Facade/立面图

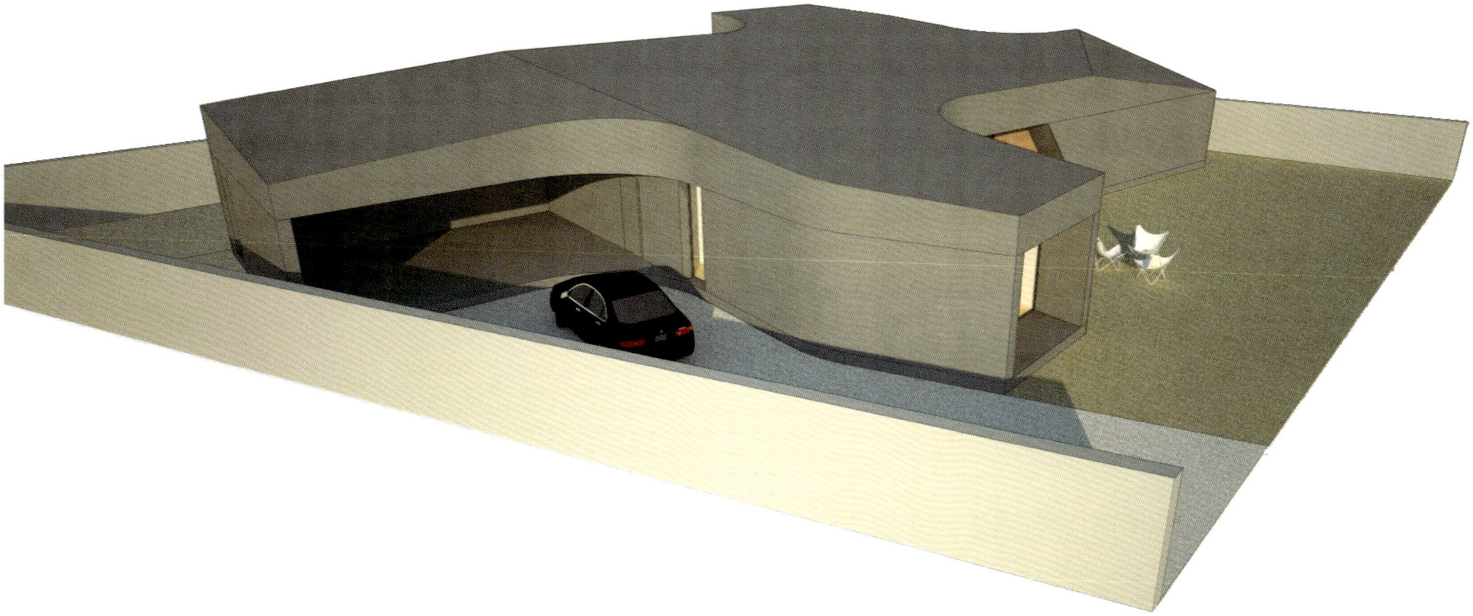

GREAT WISDOM IN SMALL SPACE

小空间 大智慧

The Green Design of Small-sized Architecture

小型绿色建筑设计实践

公共设施

PUBLIC FACILITY

The Changing Room

UNStudio

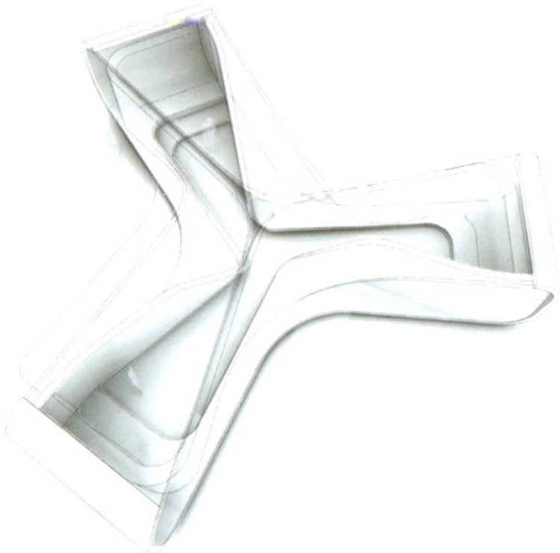

■ Architecture Design/建筑设计: UNStudio ■ Design Team/设计团队: Ben van Berkel, Caroline Bos with Christian Veddeler and Hans-Peter Nuenning, Steffen Riegas ■ Location/地点: Venice Biennale
■ Area/面积: gross floor surface 250 m^2 ■ Completion/建成时间: 2008 ■ Photograph/摄影: Christian Richters

For this year's international architecture biennale in Venice UNStudio has produced 'The Changing Room' as part of the exhibition 'out there: architecture beyond building' curated by Aaron Betsky.

The UNStudio installation in the Arsenale explores the transformative potential of the material world. Just like clothes designers, architects offer alternate looks and identities, age and income-appropriate shells. These constructions consist of a miscellaneous package of endogenous and exogenous values; things and ideas that inherently belong to architecture and its traditions, and things and ideas that do not, but that nevertheless profoundly influence architecture. How to deal with this? Can architecture still have autonomy? According to UNStudio the lesson is to 'switch it on, switch it off'... to find autonomy in brief moments of liberation.

The installation structure shows an architecture that is as supple as textile, in which floors, walls and ceilings flow into each other. On the inside, the visitor encounters a kaleidoscopic world of people posing, inviting voyeurism, and seeking transformation in their own conceptualizations of the changing room.

在威尼斯国际建筑双年展上，UNStudio创作的Changing Room项目作为参展品也出现在展会上。Aaron Betsky提出了他们的理念：建筑风格高于建筑物本身。

UNStudio在军械库的设计建筑，探索的是物质世界中的改造潜力。建筑师如同时装设计师一样，提供不同造型、个性并与不同阶层的人相匹配的各色建筑，以满足不同层次的需求。这些建筑包括了各式各样的价值观，也就是建筑及其传统固有的属性和理念，以及原本不属于建筑范畴但深刻影响建筑的那些因素。如何应对这种情况？建筑依然拥有自主权吗？UNStudio告诉我们要在短暂的自由时间里寻找自主感。

该建筑体现了一个建筑的地板、墙壁和顶棚交融在一起的效果，如纺织品一样柔软。在建筑内部，参观者仿佛遇到一个万花筒般的世界：人们摆着姿势，在这个变幻空间里，以自我的方式，寻求转变。

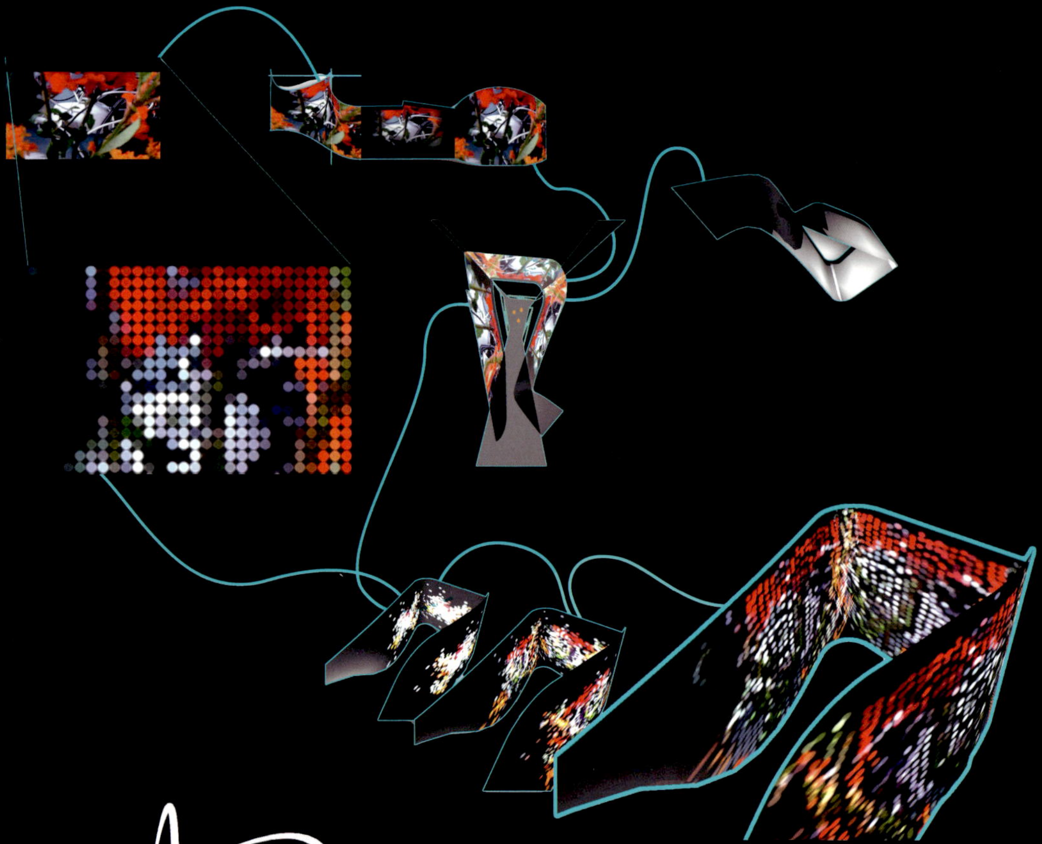

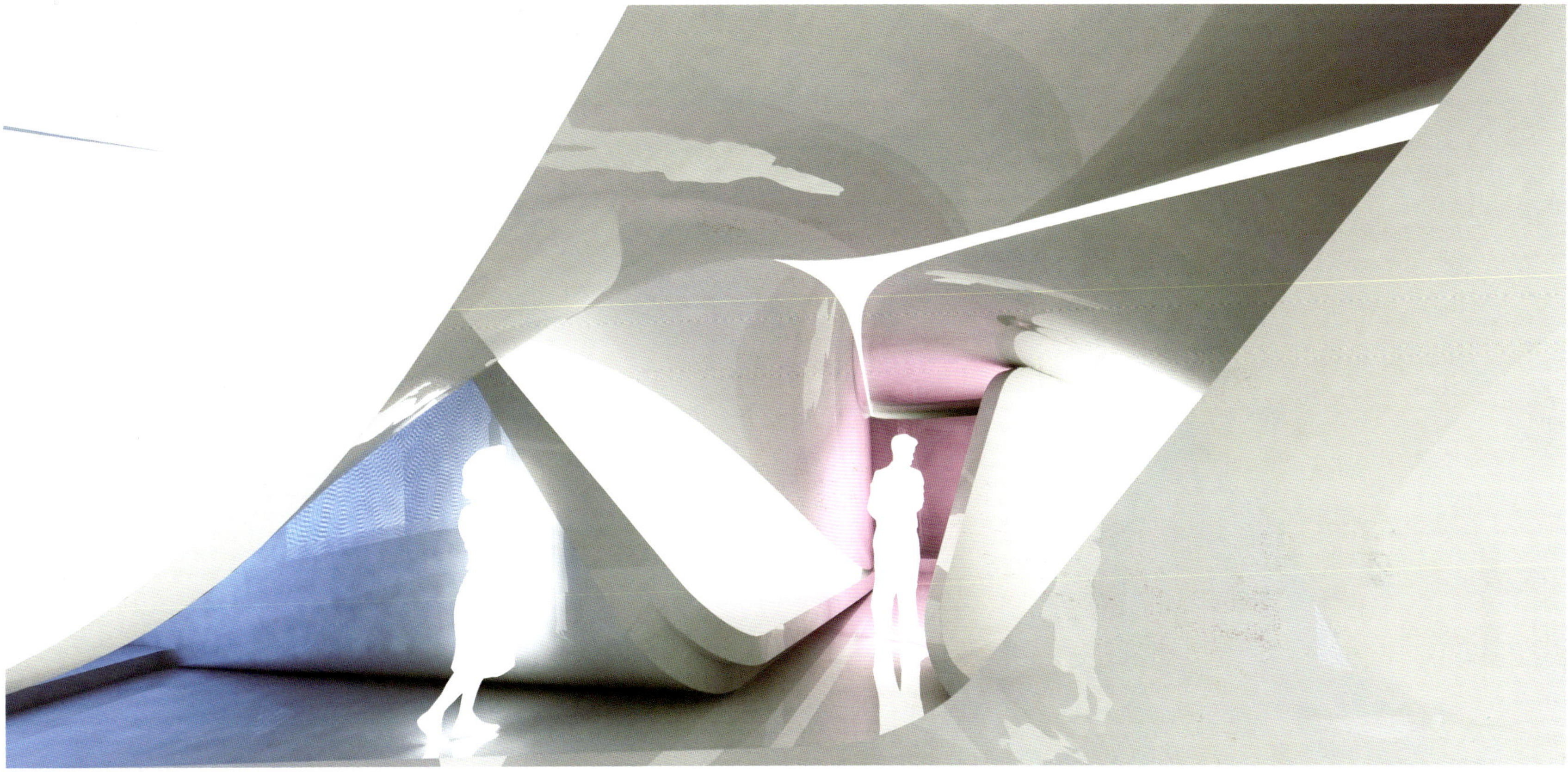

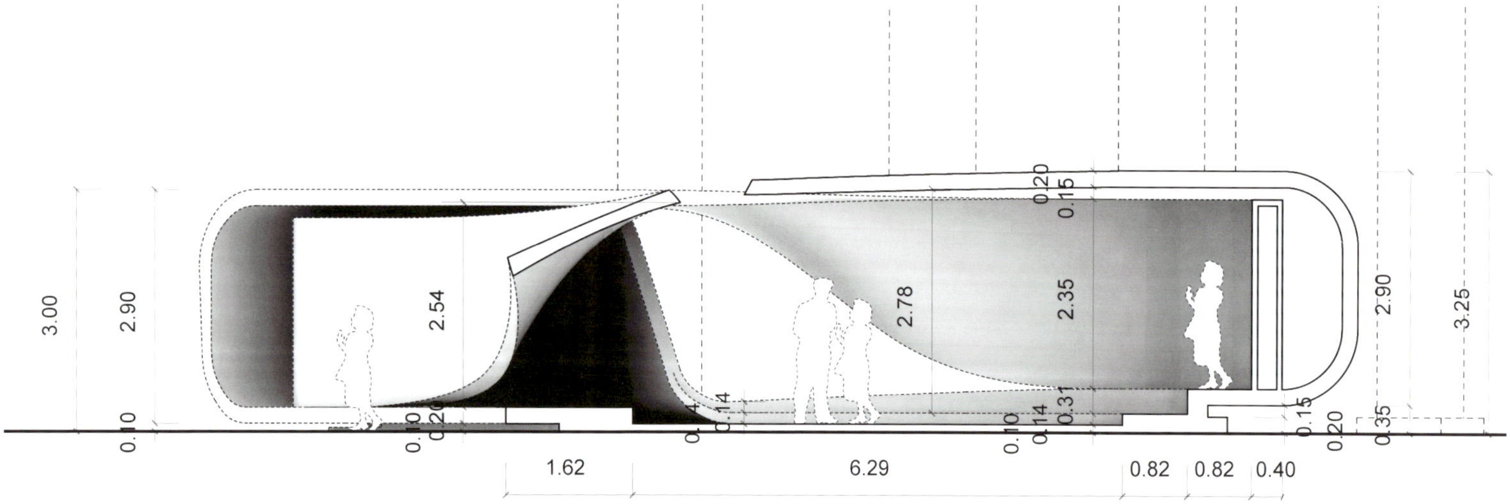
3.00
2.90
2.54
2.78
2.35
2.90
3.25
0.20
0.15
0.10
0.14
0.31
0.15
0.20
0.35
1.62
6.29
0.82
0.82
0.40

Saint-Gobain Playhouse

C. F. Møller Architects

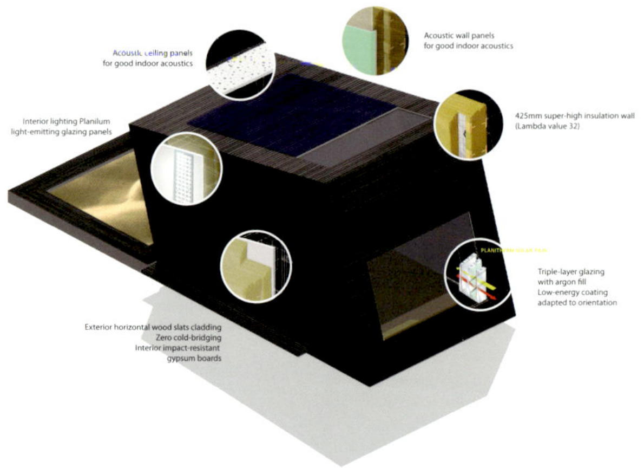

■ Architecture Design/建筑设计: C. F. Møller Architects ■ Project Architect/项目建筑师: C. F. Møller Architects ■ Area/面积: 12 m^2 ■ Completion/建成时间: 2009 ■ Photograph/摄影: C. F. Møller Architects

The sustainable playhouse designed for the manufacturer Saint-Gobain Isover, is conceived as a playhouse for children and adults - or as an unusual guest annex. The aim is to highlight the construction of energy-saving homes, and although small, the house displays details and materials which will be a part of tomorrow's solutions.

Playhouse is 4,5 meters long and 3 meters tall, and has a distinct leaning profile, with large "intelligent" glazing as windows, skylight and solar panel, as well as for interior lighting using a new technology. The walls contain a special, highly insulating glass wool, and the triple-layer energy glazing is equivalent to modern passive-house standards.

The house is so well insulated that a few candles suffice for heating, even in winter. The skewed form not only looks interesting, it also helps the building's south façade shade itself from the summer sun.

The playhouse was exhibited at Højbro Square in Copenhagen during the climate summit from the 7th to the 18th December.

为制造商Saint–Gobain Isover设计构想的可持续性玩具屋，同时满足儿童和成人的需求，也可作为招待不寻常客人的场所。建筑设计的目的是凸显建筑的节能结构，虽然房子很小，但是采用的具体方法和材料将成为未来设计方案的一部分。

这个建筑有5米长，4米宽，3米高，外表看起来是倾斜的，使用了大型“智能”玻璃窗、天窗和太阳能电池板，室内照明也使用了一种新技术。建筑的墙壁含有一种特殊的、高度保温的玻璃纤维。其三层节能玻璃窗与现代被动式节能屋的功效相当。

这个建筑的保温性非常好，以至于点燃几根蜡烛就可满足冬天取暖的需求；该建筑倾斜的造型不仅看起来有趣，而且也可以帮助建筑的南墙面自行遮挡烈日的阳光。

该玩具屋是在2009年12月7–18日在哥本哈根气候峰会期间Højbro广场上展出的。

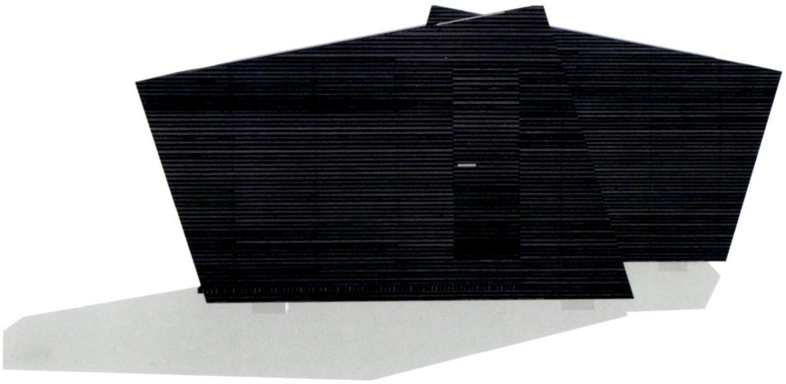

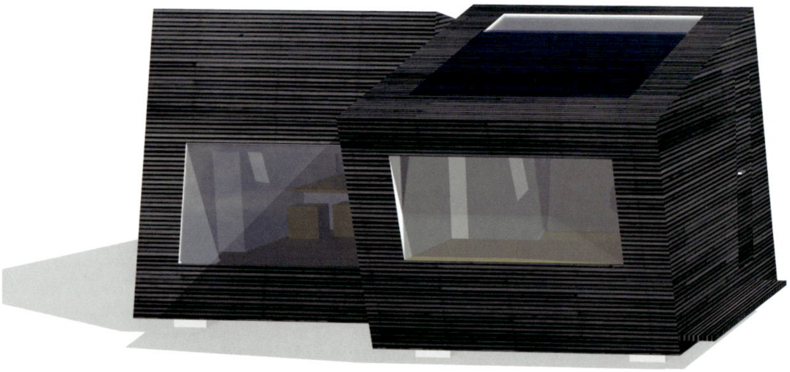

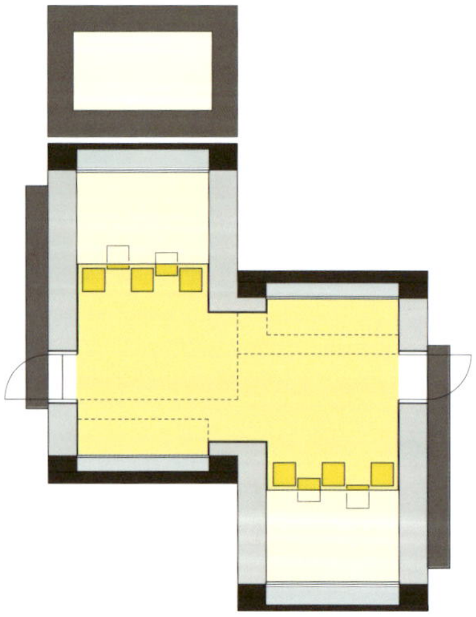

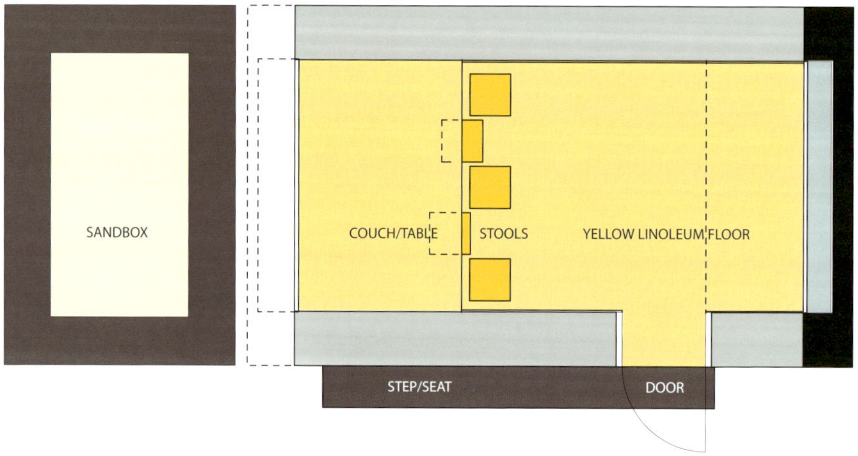
SANDBOX
COUCH/TABLE
STOOLS
YELLOW LINOLEUM FLOOR
STEP/SEAT
DOOR

Plan/平面图

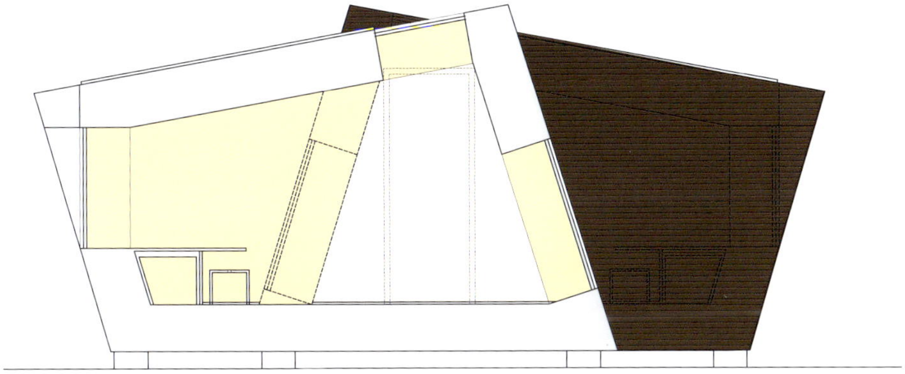

Section/剖面图

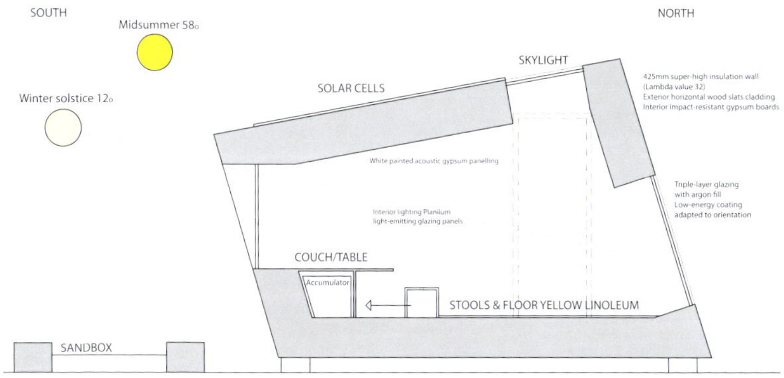

Section/剖面图

Tree Hotel

Tham & Videgård Arkitekter

- Architecture Design/建筑设计: Tham & Videgård Arkitekter
- Project Architect /项目建筑师: Martin Videgård, Bolle Tham, Mia Nygren, Andreas Helgesson, Julia Gudiel Urbano
- Location/地点: Harads, Sweden
- Area/面积: 16 m^2
- Completion/:建成时间: 2010

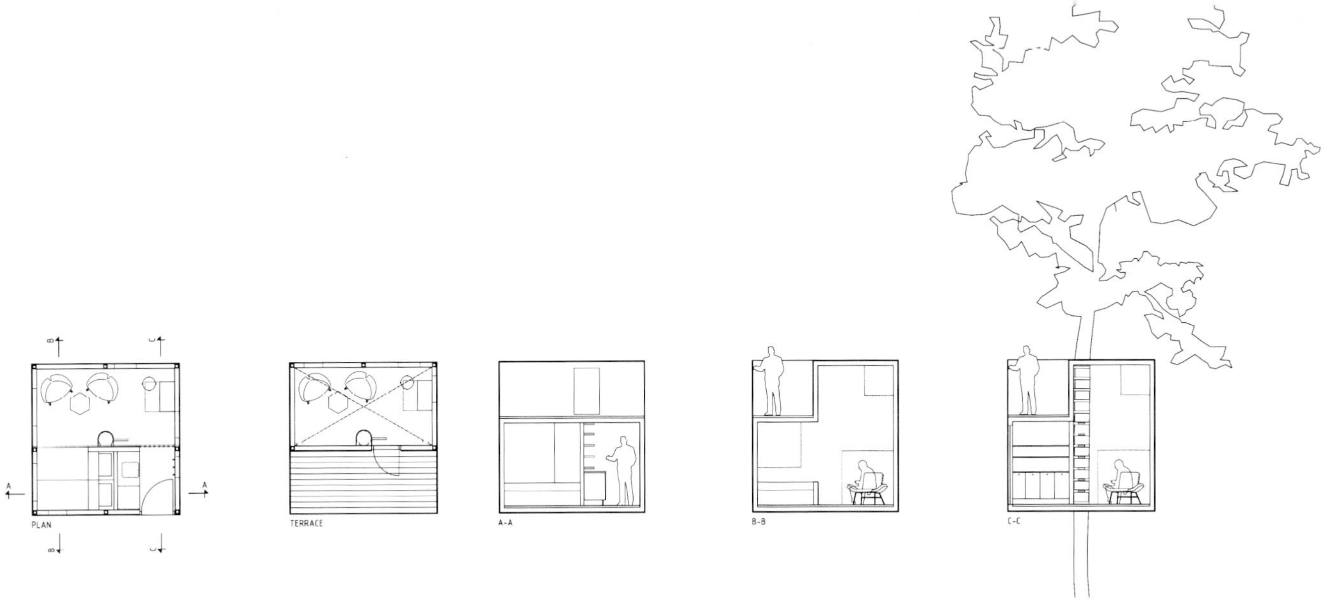

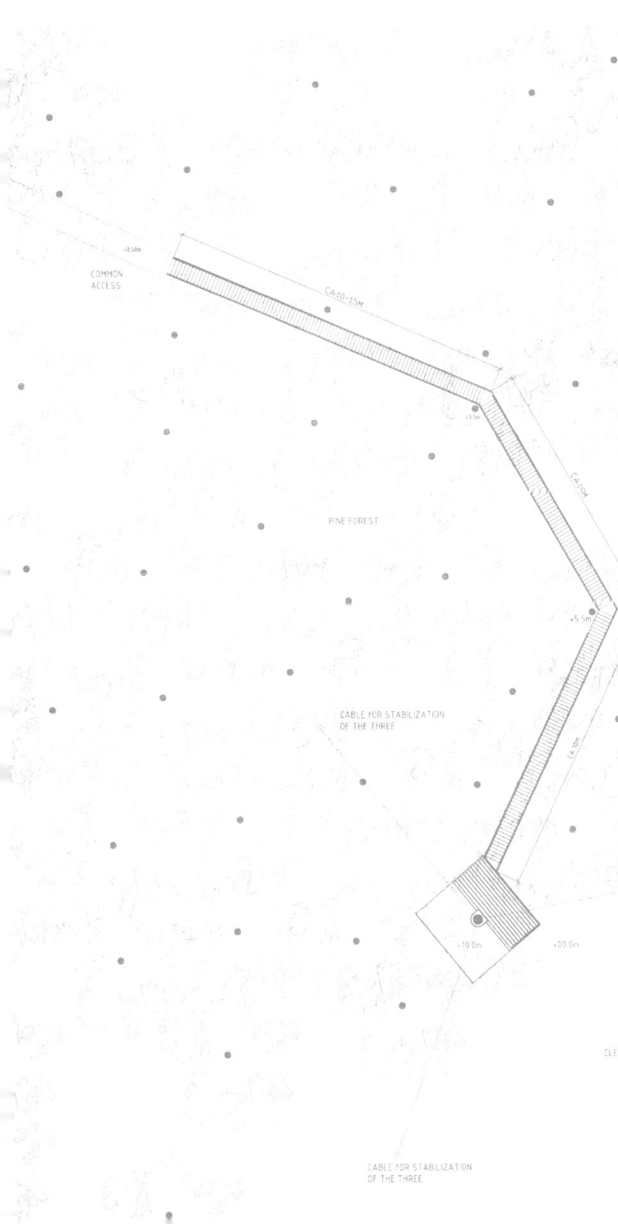

A tree hotel in the far north of Sweden, near the small village of Harads.

The starting point is to create a shelter up in the trees; a lightweight aluminium structure hung around a tree trunk, 4x4x4 meters boxes clad in mirrored glass. The exterior reflects the surroundings and the sky, creating a camouflaged refuge. The interior is all made of plywood and the windows give a 360 degree view of the surroundings.

The tree hotel concept meets the increasing interest in wild life eco-tourism where one encounter and experience the pristine nature of Sweden.

The functions included provides for a living for two people; a double bed, a small kitchenette and bath, a living room and roof terrace. Access to the cabin is by a rope ladder or a rope bridge connected to the next tree.

To prevent birds colliding with the tree hotel, a transparent ultraviolet colour, visible for birds only, will be laminated into the glass panes.

The tree hotels will be managed by the Brittas Pensionat (a small hotel nearby) that will also cater for and offer additional services and support. The project is currently in progress and the client, Brittas Pensionat, is planning to have the six first cabins completed in 2010.

在瑞典北部一个叫Harads小村旁边有一个Tree Hotel项目。它的最初设计想法是能够建造一个在树下避风遮雨的地方。这是一个轻薄的铝制结构，围绕着树挂在空中，是一个长、宽、高均为4米的小屋。外表是一层镜面玻璃，这样可以将周围的景物和蓝天都反照出来，遮掩了这个避难所；其内部是由胶合板构成，可以享受周围360度的风景。

Tree Hotel的理念与当下流行的野外生态旅游相结合，可以让游客亲身感受原始状态下的瑞典自然生存环境。

这个小屋是为两个人设计的，一张双人床，一个小的厨房，一个浴室，一个客厅和屋顶平台。小屋的出口是软梯或与旁边的树木连接在一起的绳桥。

为了防止鸟类动物与Tree Hotel碰撞，外玻璃表面叠压了一层能够发出只有鸟类能看到的透明色紫外线。

Brittas Pensionat是附近的一家酒店，将负责管理Tree Hotel，他们同样会为Tree Hotel客人提供餐饮服务和其他额外的服务和帮助。这个项目目前还在进行中，Brittas Pensionat计划在2010年完成最初的六个小屋。

HOTO FUDO

TAKESHI HOSAKA ARCHITECTS

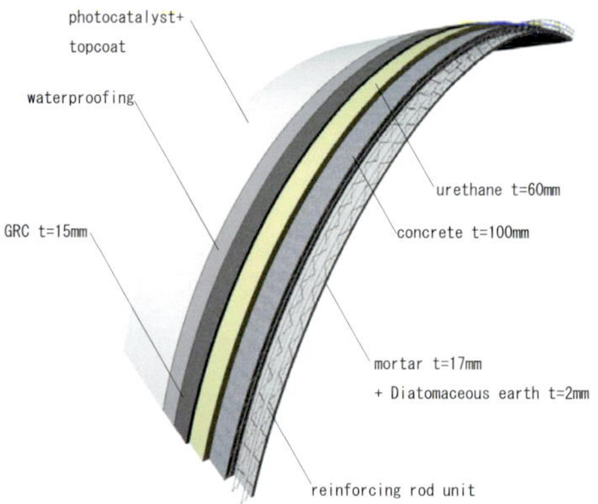

■ Architecture Design/建筑设计: TAKESHI HOSAKA ARCHITECTS ■ Project Architect/项目建筑师: Takeshi Hosaka ■ Location/地点: Yamanashi, Japan ■ Area/面积: site 2,493.82 m^2/building 733.98 m^2/floor 726.30 m^2 ■ Completion/建成时间: 2009 ■ Photograph/摄影: Koji Fujii / Nacasa&Pertners Inc.

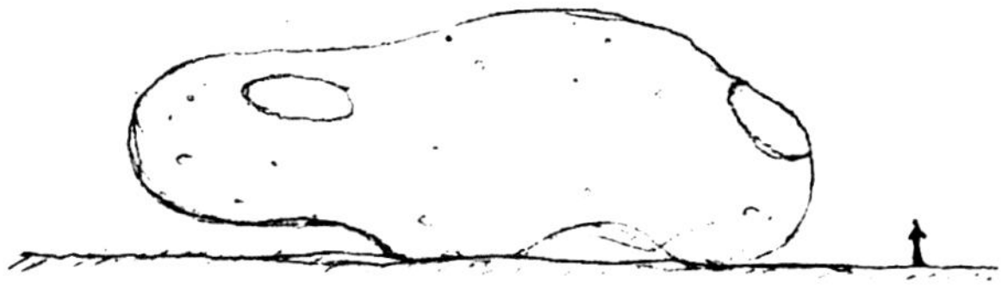

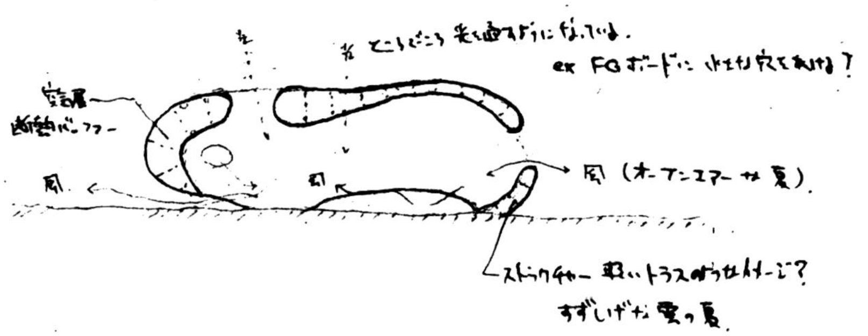

This is a building like inside and outside.

The project was planned on the site with Mt. Fuji rising closely in the south and the two sides facing the trunk roads.

This building seems to belong to such nature objects as mountains and clouds. It is made from soft geometry, which will not arise from the figures like quadrangles and circles. By continuously operating innumerable polygon mesh points, we have determined the shape that clears the conditions such as the consistency as shell construction and the undulations that ward off rainwater in spite of its free geometry. The RC shell with cubic surfaces creates such spaces as 530 square meters of seats, 140 square meters of kitchens, and 50 square meters of rest rooms, in such a manner that it envelops and opens them.

This building has no air conditioners. It is open to the air at most seasons, and people have a meal in the air like outside air. The curved acrylic sliding door is closed only during the strong wind and the coldest season. Giving 60 mm thick urethane insulation to the outside of the RC shell and keeping a stable RC temperature secures a stable temperature environment for the building like inside and outside, and also reduces the deformation volume due to the temperature of RC to make the building last longer. For the lighting plan, we have determined such illumination as makes people simply feel changes in the evening light and does not make insects gather around the lights. When it rains, rain comes in near windows and doors. In the spaces where rain does not come in, people enjoy the sound of raindrops. When it is foggy, the fog comes into the building. When it snows, it becomes a landscape buried in snow, and birds and animals will visit there. In this place like the middle between nature and art, people eat hoto rich in natural ingredients.

HOTO FUDO项目包含着既室内又户外的理念。

该项目建在紧邻富士山南坡的地方，建筑两侧都面向主干道。

该建筑与自然浑然一体，像一座山或是朵朵云彩。它由线条柔和的几何造型构成，既不是四四方方，也不是圆形的球体。设计师不断地运用大量的网点，最终确定了其形状，并确定了各项设计参数，比如贝壳结构的坚实度，还有其波动的曲线。虽然是自由的几何图形，但它也要具备遮挡雨水的功能。这个530平方米的空间是由钢筋混凝土的贝壳与三层曲面结合而成的，其中厨房面积约为140平方米，洗手间面积约为50平方米，它既封闭又开阔。

在这个建筑里，没有安置空调，因为一年四季大部分时间，这里都处于通风的状态，因此，人们在进餐时微风轻拂，仿佛是在户外野餐一般。只有在强风或是最冷的季节才会关闭弧形的丙烯酸拉门。在钢筋混凝土贝壳的外层安装60毫米厚的聚氨酯保温板，稳定其温度，保证建筑物室内的温度与室外接近，同时也减少了温度对建筑物造成的损坏，延长了使用寿命。照明设置，既可以让人们感觉到夜晚灯光的变化，又不会吸引昆虫围绕在照明设备附近。下雨时，雨水会滑落在玻璃和门上，人们在室内可以尽情聆听雨声。在起雾的天气里，雾气会弥散到室内。当下雪的时候，这个建筑就会变成埋在雪下的风景，很多小动物也会光临此地。这里仿佛是一个介于自然与艺术的中间站，人们在自然环境中欣赏着艺术。

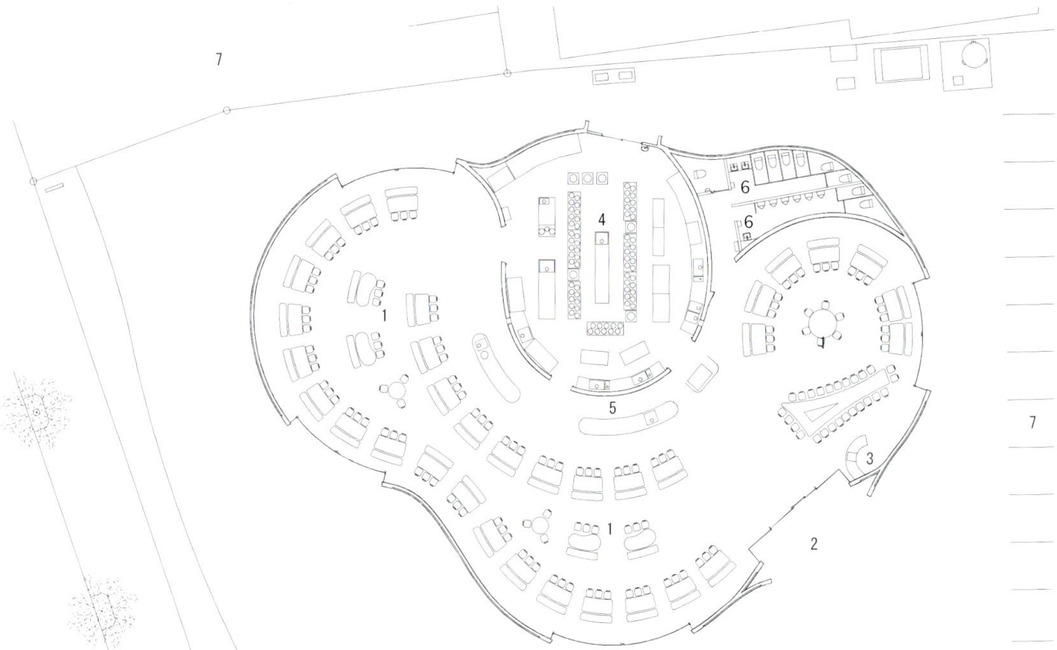

First Floor Plan/一层平面图

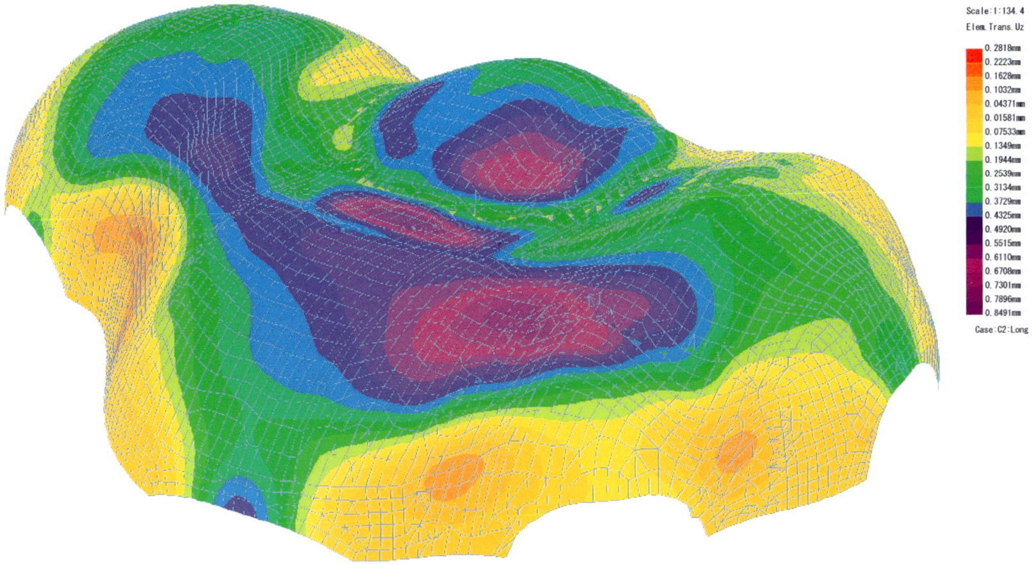

Scale:1:134.4
Elem. Trans. Uz
0.2818mm
0.2223mm
0.1628mm
0.1032mm
0.04371mm
0.01581mm
0.07533mm
0.1349mm
0.1944mm
0.2539mm
0.3134mm
0.3729mm
0.4325mm
0.4920mm
0.5515mm
0.6110mm
0.6708mm
0.7301mm
0.7896mm
0.8491mm
Case:C2:Long

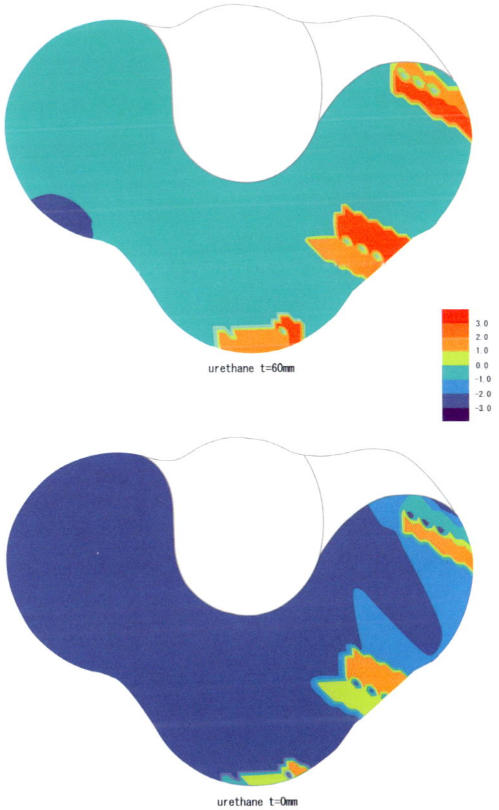

urethane t=60mm
3.0
2.0
1.0
0.0
-1.0
-2.0
-3.0
urethane t=0mm

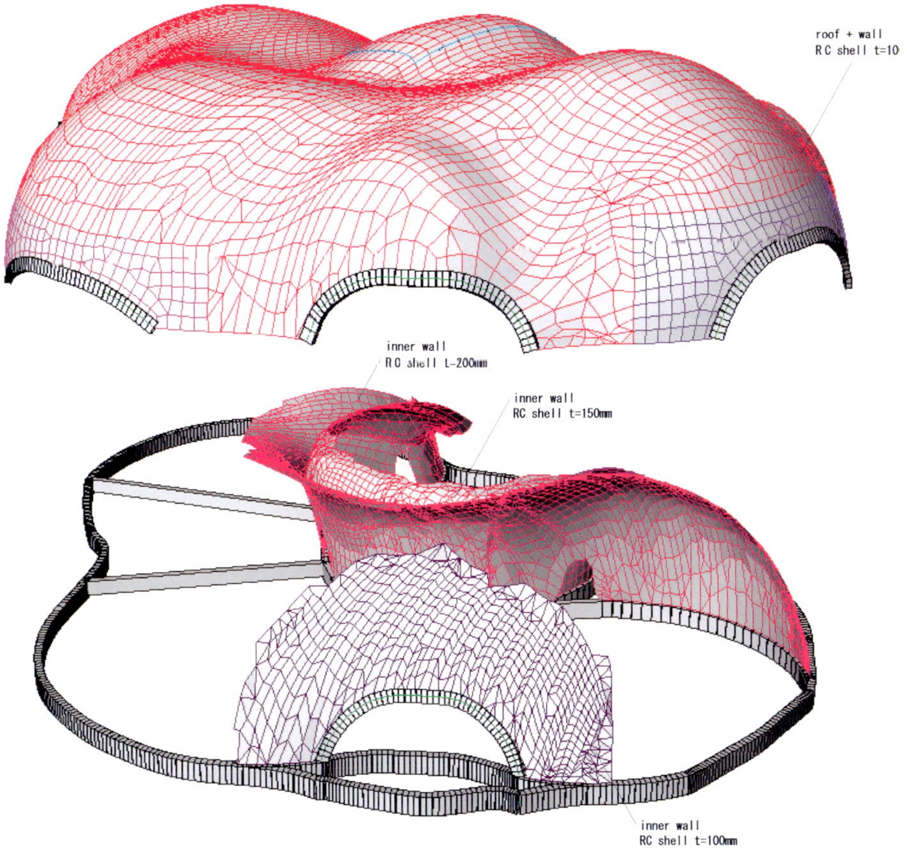

roof + wall
RC shell t=10
inner wall
RC shell t=200mm
inner wall
RC shell t=150mm
inner wall
RC shell t=100mm

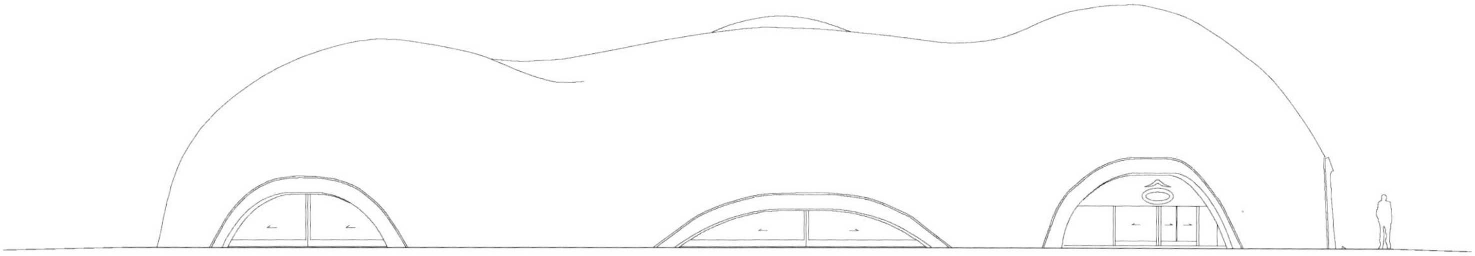

Facade:West/西立面图

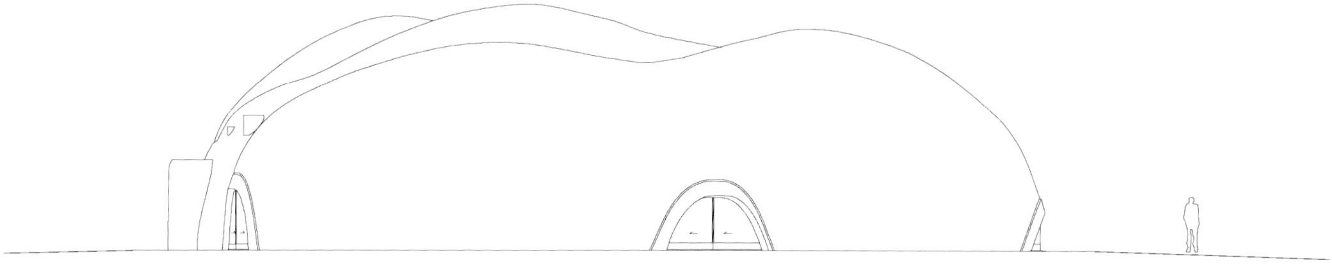

Facade:North/北立面图

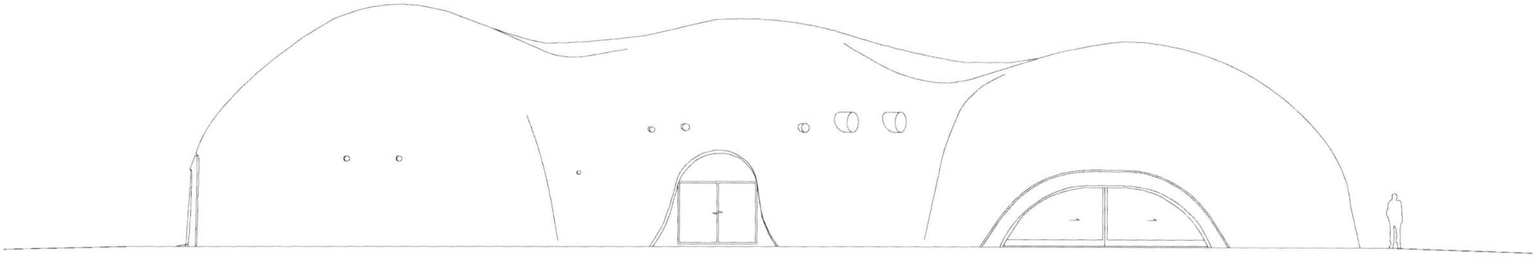

Facade:East/东立面图

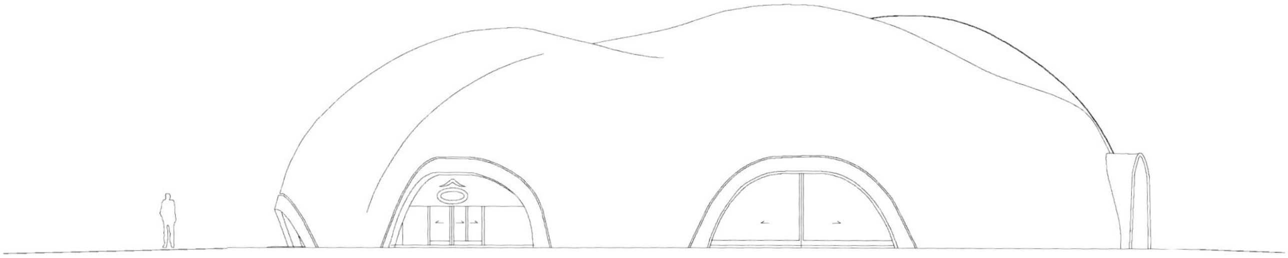

Facade:South/南立面图

Hongodai Christ Church school & nursery

TAKESHI HOSAKA ARCHITECTS

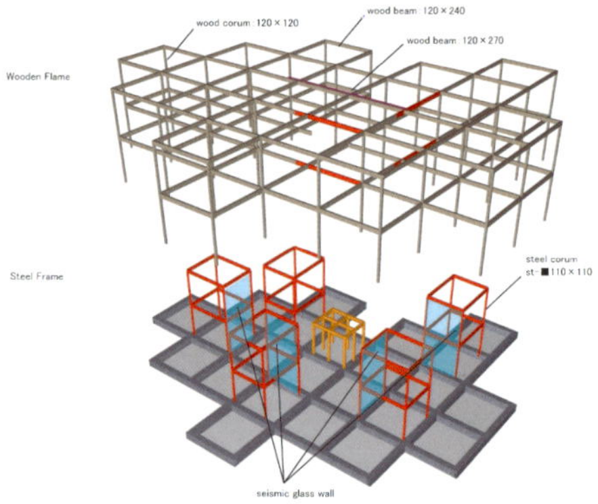

■ Architecture Design/建筑设计: TAKESHI HOSAKA ARCHITECTS ■ Project Architect/项目建筑师: Takeshi Hosaka ■ Location/地点: Yokohama, Japan
■ Area/面积: site 1,332.92 m^2/building 265.97 m^2/total floor 531.94 m^2 ■ Completion/建成时间: 2010 ■ Photograph/摄影: Masao Nishikawa

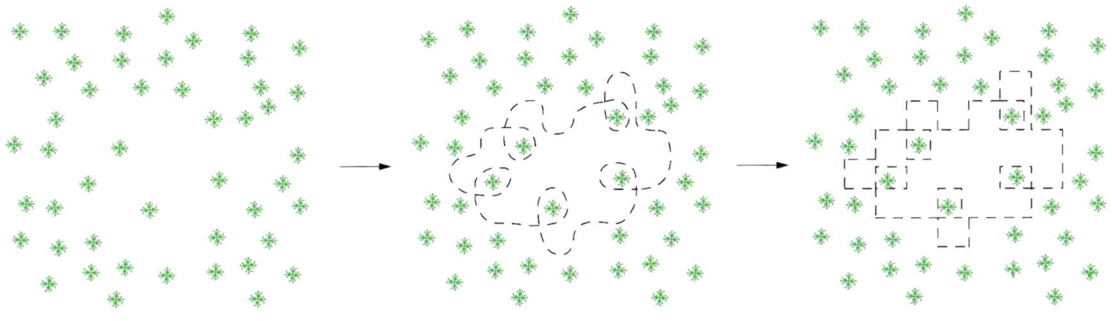

It is a complex facility made up of a private school and a nursery operated by a protestant church.

The site is in a green area of Yokohama, surrounded by playgrounds, parks and small hills. The plan is to blend a place for children ranging from 3rd year high school students with the green forest.

The line drawn with one stroke run through the forest while dodging and surrounding trees. It is placed in the forest as a simple grid by wood frames. As the building in which the children spend time, steel frames in the five courts play a part in providing better earthquake resistance, in the wooden construction of transparent grid with wooden frames, wooden floors and wooden sashes. Sliding glass doors are the interfaces between the inside and outside. Sunshine filters through the trees into the building and the breeze wafting through the forest also wafts through the building. Being inside the building is just like being in the forest and the exterior of the building also seems to belong in the building, making it possible for children to act as they like.

Morning service is held in the hall on the second floor, and then the children go to their own places. The children in nursery run through the forest and eat their lunches under the trees. Lessons are held under the sunlight filtering down through the trees and picture books are read in the forest.

This architecture is representing a new idea that how we should build in forest. It is not too artificially controlled, I hope this broad-minded architecture between the natural and the artificial will be the place where children can feel free and be themselves.

这个复合式的建筑是由新教派教堂管理下的私立学校和幼儿园组成的。

学校位于横滨的一块绿化区域，周围有游乐场、公园，还有些小的丘陵。客户计划将此地设计成为供孩子们与高中生一起娱乐的绿色森林。

设计师贯穿森林划出一条线，使学校和周围的树木分开。这栋房子建在森林中，木质的框架让它看起来像一个简易的格栏，考虑到它是供孩子们使用的，设计师特意增强了建筑物的抗震性，分别在五个庭院部分使用了钢结构。建筑物从里到外都是透明的木质建筑格栏，木质地板以及木质窗扇，滑动的玻璃门窗就是一个分隔内外的分界面。阳光穿过树枝洒在室内，微风掠过树丛飘进房间。在室内时，人们仿佛置身于丛林之中，在户外时，又恍如身在室内一般。这样一来，孩子们在这里就可以完全放松，尽情玩耍。

早上活动的区域设置在二层的大厅，孩子们活动后会回到各自的座位上。中午幼儿园的小朋友们会穿过树林在树下午餐，下午的课程也会在树林里继续，午后的阳光透过树枝细碎地洒在孩子们的画册上。

这个建筑设计代表了一种新的理念即如何在丛林中建造房屋。这种建筑不完全是由人控制的，设计师希望这种人工与自然结合的魅力建筑能够成为孩子们自由成长的地方。

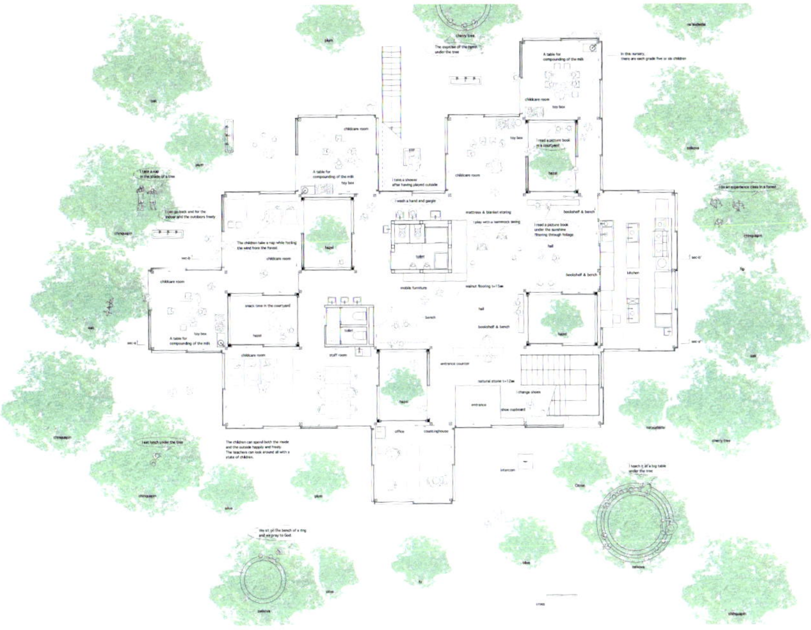

First Floor Plan/一层平面图

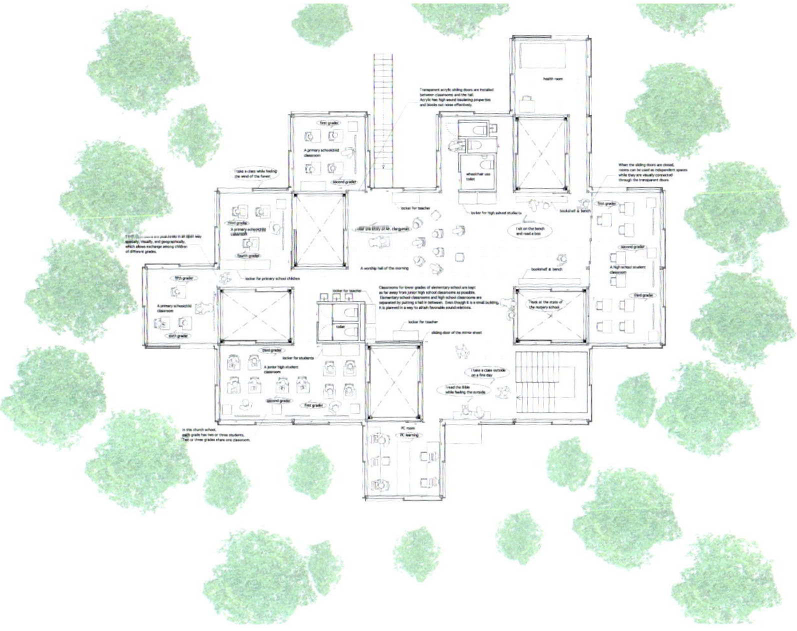

Second Floor Plan/二层平面图

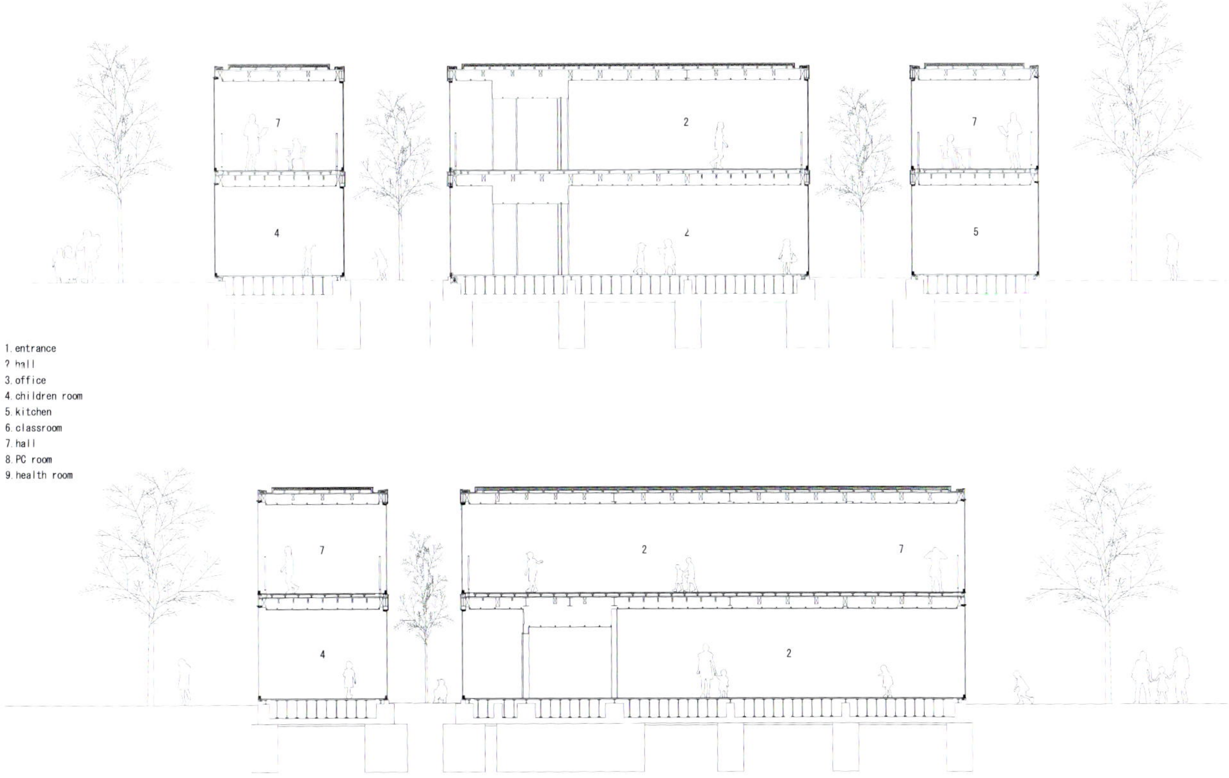

Section/剖面图

Hiroshima Park Restrooms

Future Studio

■ Architecture Design/建筑设计: Future Studio ■ Project Architect/项目设计师: Bunzo Ogawa ■ Location/地点: Hiroshima, Japan ■ Area/面积: A type 15.56 m², B type 11.88 m², C type 7.62 m²
■ Completion/建成时间: 2009 ■ Photograph/摄影: Toshiyuki Yano

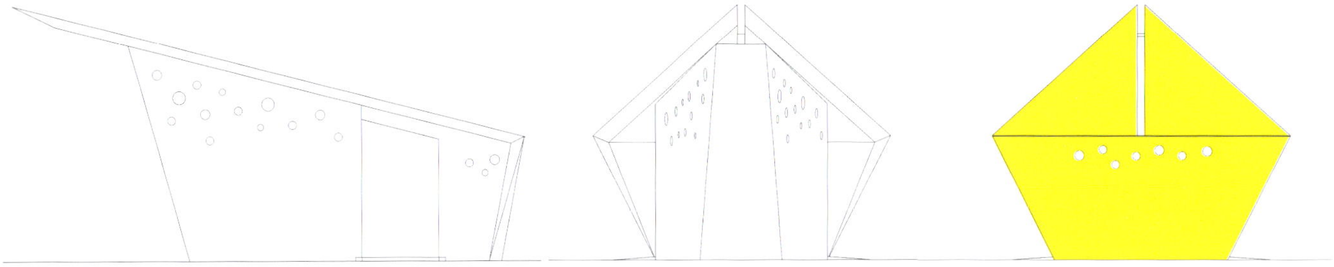

This is a restroom project build in the park of Hiroshima city, which was adopted by a design competition. It is the unique public project that approximately five restrooms are mass-produced in parks per year as regular design of the city.

It is "interaction of man and a city" that I aimed. An absolute axis as gravity exists on the earth, and the horizontal and vertical axes are assured. I aim to make an absolute axis in the city by being embedded the direction in infrastructure building in the city.

Necessary space on the function of this rest room was for multipurpose restroom in which a circle of 1.5m can be inscribed, a Japanese style restroom booth, a urinal, a washbowl, passage of effective 1.2m. The north-south wall is inclined to create spacious space. In addition, I incline the roof, the north side is high, the south side is low, to be able to look at the roof facing north from eye-level. This restroom project assumes a mass production. Therefore it should be constructed in any sites of any parks with the tip of triangle roof pointing at north. Accordingly I provide with 2 entrances of east and west side, so that the plan is able to turn the other way around as it functions whichever entrance.

I designed 3 variations of plans in this project. It is possible to build the basic plan (A type) in about 80% of parks in Hiroshima-city. But smaller plan was required for small parks to make them fit in with building coverage 2% in the park set by law. It is the B type except a Japanese style restroom booth from A type, and the C type only for the multipurpose restroom.

该项目是要在广岛市公园修建公共卫生间，设计方案是通过设计比赛选出的。这是个特殊的公共项目，特殊在每年该市都会在园区内修建约5个公共卫生间，以此作为该市城市规划设计的一部分。

设计师力求该设计达到“人与城市互动”的效果，并希望通过这些公共基础建筑，在该市建造出一个绝对的坐标轴。

出于卫生间多功能的考虑，设计师在卫生间内所需的空间范围内，安放了一个日式的厕盆、一个小便器、一个面盆并设计一个1.2米的通道。其中，南北向的墙体是倾斜的，从而创造出更多的空间。另外，设计师将屋顶设计为倾斜的，北面高，南面低，人们面朝北望去，能清楚看到该建筑物的屋顶。鉴于该卫生间利用率较高，因此，在任何一家公园的任何角落修建该建筑物时，都要一致地修建三角形屋顶，指向北面。设计师也相应地设计出两个入口位置，两个入口分别在东、西两侧，这样，设计就实现了从东、西两个入口均可进入的构想。

该项目有3个不同的设计方案。在广岛市约80%的园区，都可以采用基本方案（A类），在一些小的公园里，为满足法律中关于建筑占地面积不超过园区面积2%的规定，应采取更小型的设计方案，此时可以采用B类设计方案，与A类设计方案相比，它只少了一个日式的厕盆，而C类方案仅适用于多功能卫生间的建造。

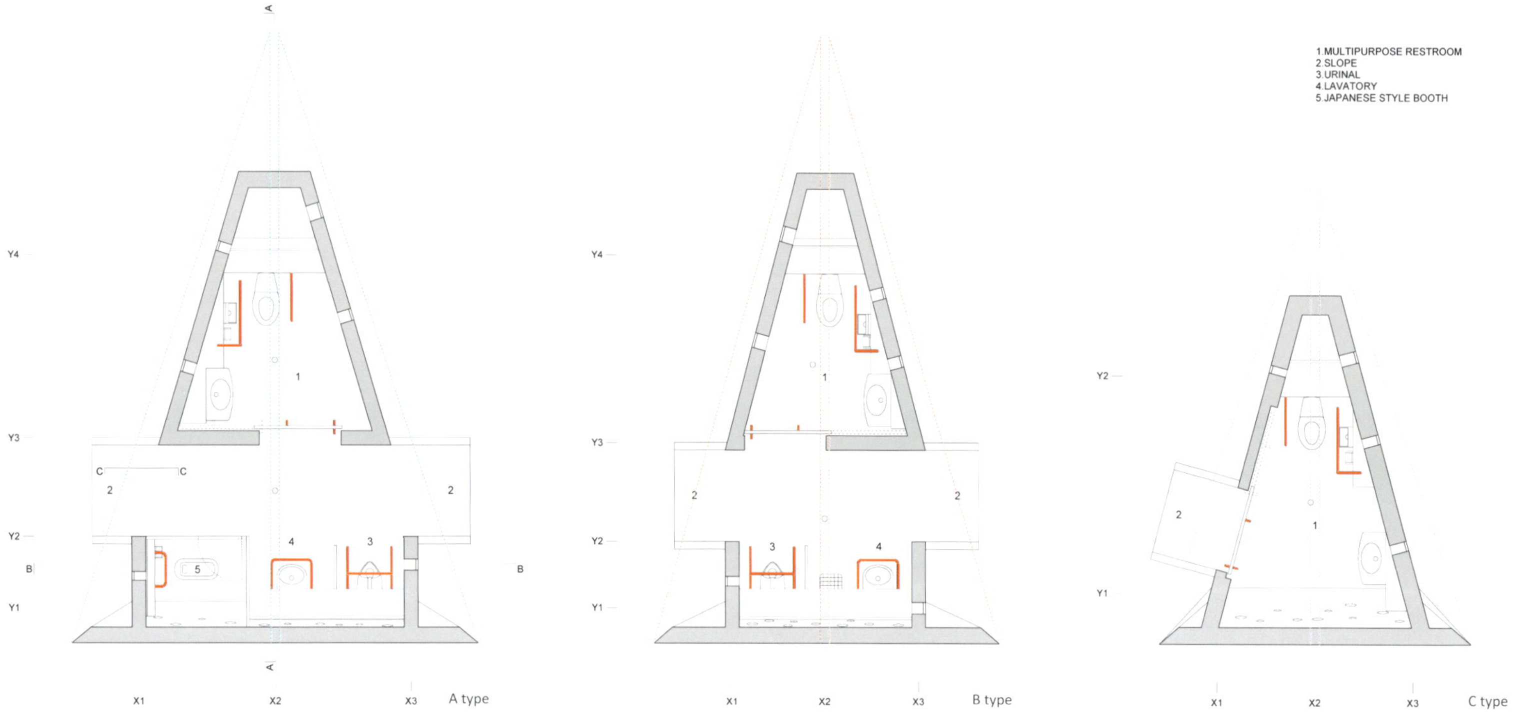
1.MULTIPURPOSE RESTROOM
2.SLOPE
3.URINAL
4.LAVATORY
5.JAPANESE STYLE BOOTH
Y4
Y3
Y2
Y1
X1
X2
X3
A type
B type
C type

Plan/平面图

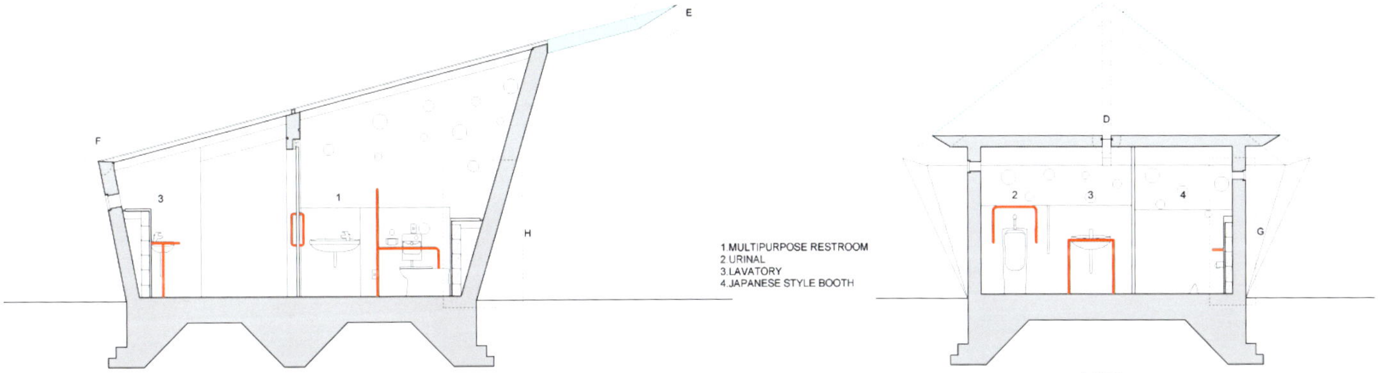
E
F
3
1
H
1.MULTIPURPOSE RESTROOM
2.URINAL
3.LAVATORY
4.JAPANESE STYLE BOOTH
D
2
3
4
G
A type

Section/剖面图

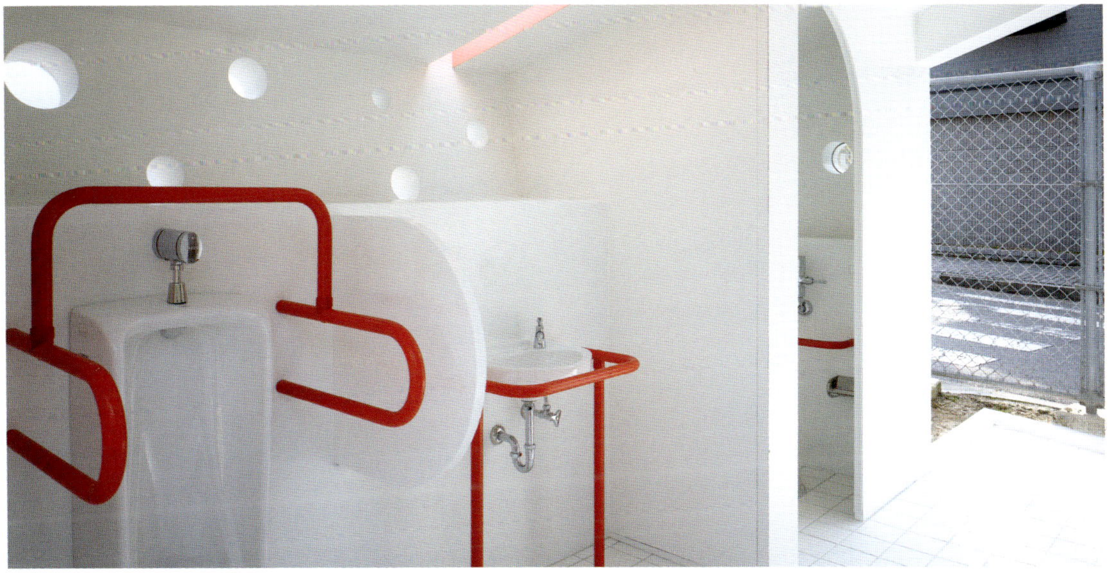

Burnham Pavilion

UNStudio

■ Architecture Design/建筑设计: UNStudio ■ DesignTeam/设计团队: Ben van Berkel, Caroline Bos with Christian Veddeler, Wouter de Jonge and Hans-Peter Nuenning, Ioana Sulea
■ Location/地点: Chicago, USA ■ Photograph/摄影: Christian Richters

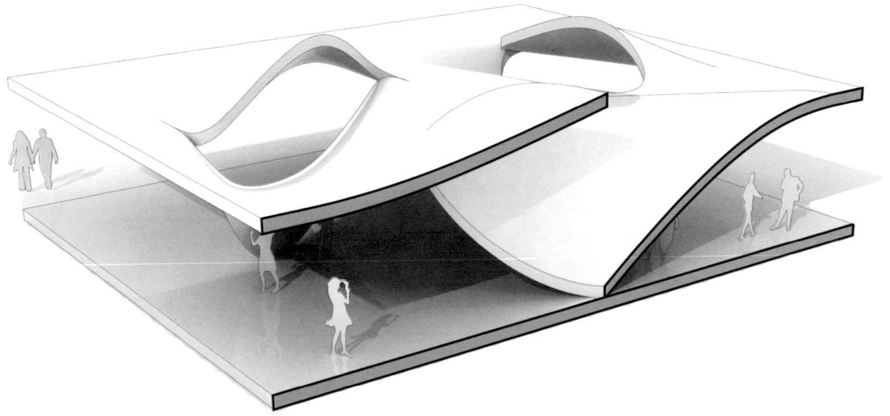

The design of the UNStudio pavilion initially uses the orthogonal setup of the city and park grid. The edges of the pavilion follow the strict grid system of the surrounding city and park geometry. Burnham's Plan one hundred years ago introduced this generic grid as the main generator of the city texture. However alongside this he also introduced a device to read both specificity and variation in the form of the diagonal boulevards, thereby creating specific vistas throughout the city.

The UNStudio pavilion operates in a similar way, but in contrast uses the entire spectrum of 360 degrees. Instead of Burnham's tectonic layering of the city, the pavilion introduces a gradient between its ingredients of floor, wall and ceiling in a floating and continuous form. The hierarchy of the horizontal or vertical plane is converted into an understanding of a space as continuous transformation and fluidity. The initially horizontal panorama on top of the pavilion's podium smoothly shifts diagonally into the three roof openings, framing vertical views of the city skyline. The ambivalence of directionality and the introduction of continuous flow in the structure allows for a smooth opening-up of spaces, directions and, importantly, the most diverse vistas from which to frame and read the city context.

UNStudio的Burnham Pavilion项目最初设计采用的是城市与公园互相交错的网络形式。展厅的外围依照严格的城市网络布局，并依据公园的几何形状设计。一百年前的伯纳姆计划就提出了这种观点：共有网络可以作为城市特质发展的主要源泉。不仅如此，他还介绍了一个方法，能够测量主要街道对角线的精准值和变化值，从而可以规划出整个城市的具体前景。

UNStudio的Burnham馆运用了类似的设计手法，在使用方面却完全不同，它的设计摒弃了伯纳姆的城市分层构造，展示了在一个漂浮不定的状态下，地板、墙和顶棚的坡度。水平或是垂直的平面被设计成一个像流水一样不断变幻和流动的空间。在展厅的顶端，全景式的水平台流畅地变换出三个屋顶开口，把城市的天际线框成垂直的视面。方向的交互性和流线形的结构展示了一个平滑的开放式空间，更重要的是，从这里可以领略到城市最多样化的风景。

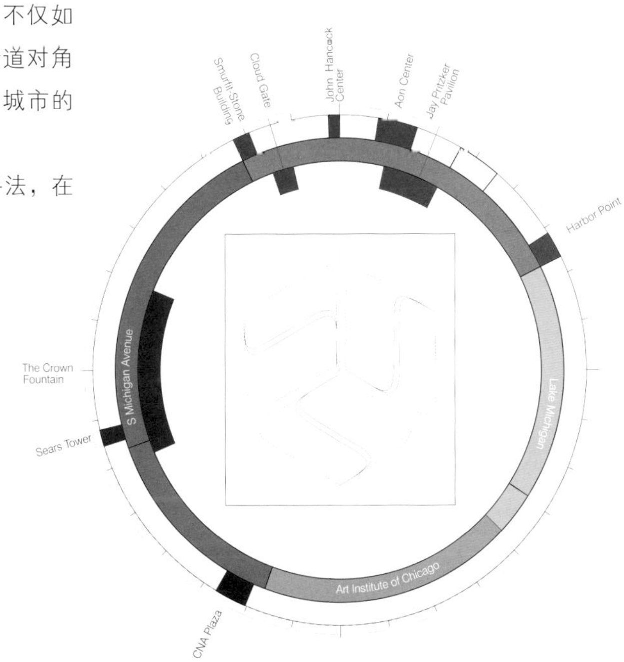

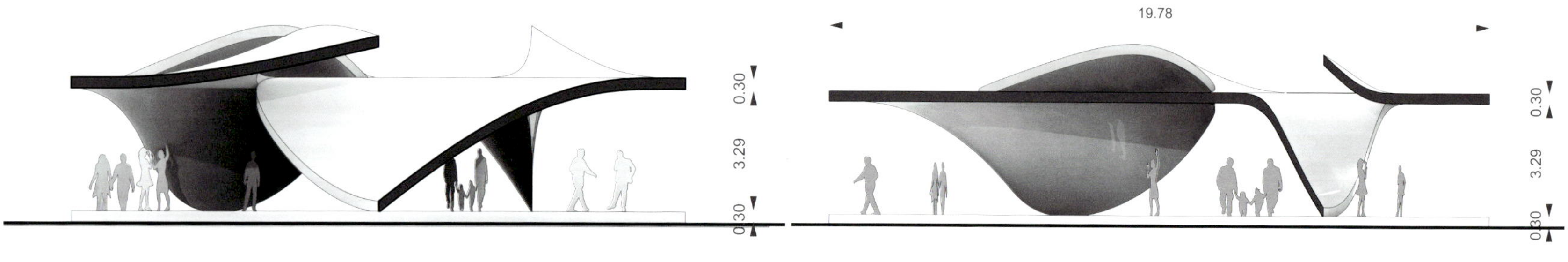

Section/剖面图

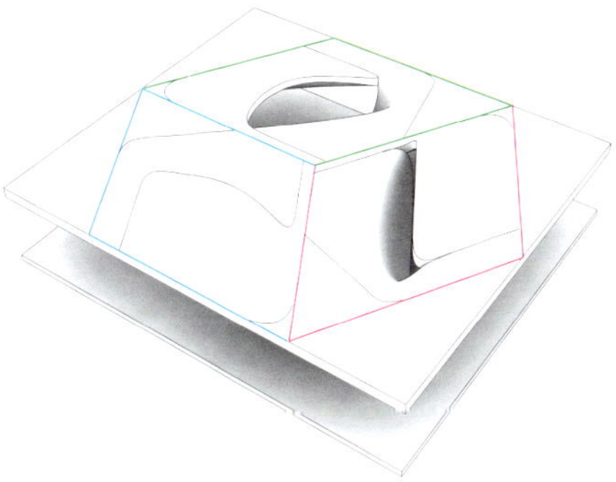

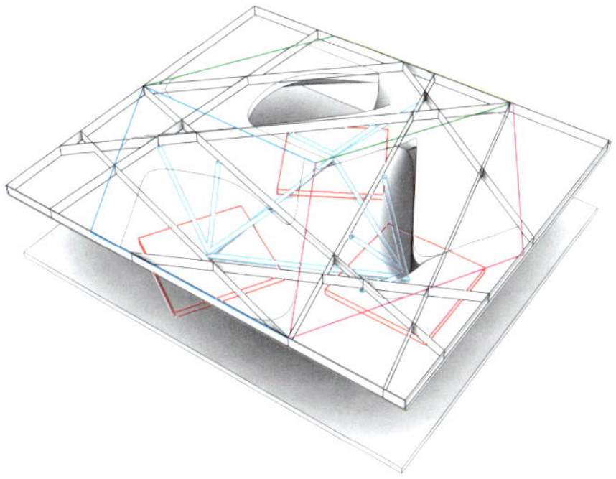

La Finca

Buscando La Aurora

- Architecture Design/建筑设计: Buscando La Aurora
- Project Architect/项目建筑师: Carlos Coronel, Hector de la Peña
- Location/地点: Morelos, México
- Area/面积: site 4,500 m² / construction 144 m²
- Completion/建成时间: 2010
- Photograph/摄影: Marcos Betanzos, BulAu

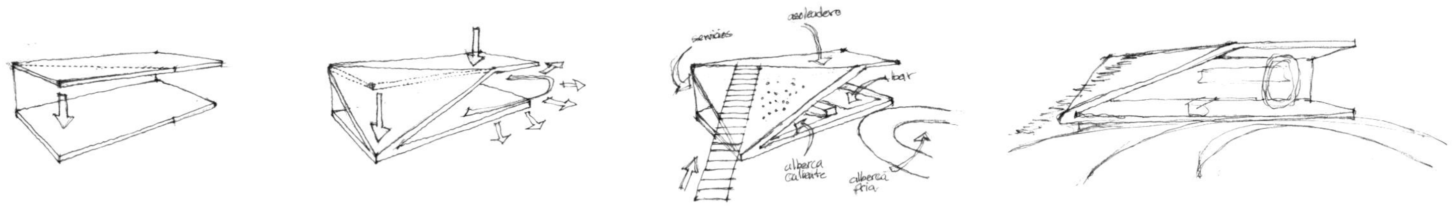

The site is located in Morelos, a small touristic state south Mexico City, in a small town called Sta. Catarina. The location is a 4,500 square meters lot, where a 35 years old weekend house sits.

The client required a program of recreation activities that included a small pool, a bar area, lounge area, bathroom service, and tanning zone.

The solution gathers all the activities together in a floating rectangle with two extracted voids of similar dimensions on the lower floor that becomes the pool and the lounge area. The floating roof above them gives covering from rain and sun and sits on a triangular volume that becomes the service block.

The rectangle shaped roof on plan becomes a diagonal when one of its vertices touches the respective one down. The sloped triangle roof holds the stairs to climb up to the top, where the tanning zone is, with incredible views towards the mountains.

本项目位于小镇圣卡塔琳娜，在墨西哥城南部的以旅游著称的莫雷洛斯州境内。项目所处的地点是一个占地面积约4 500 平方米的曾有着35年历史的周末度假村。

客户要求增加娱乐活动设施，包括增加一个小的游泳池、一个吧台区、一个休息室、一个日光浴场和洗浴设施。

该设计方案选择在一个悬空的矩形平顶上集合了所有的项目。在这个比平顶更低一层的位置上，设计师开发出了与平顶类似的两个空间，并将其作为游泳池和休息区。这个悬空的屋顶能够遮阳避雨，所构成的三角形空间则变成了配套服务区。

设计按计划进行，把这个矩形平顶沿对角线对折，形成两个彼此垂直的二角平面，在倾斜的三角形区域屋顶上安装楼梯，可以方便游客爬到顶层，让游客在享受日光浴的同时还能欣赏美不胜收的山水风景。

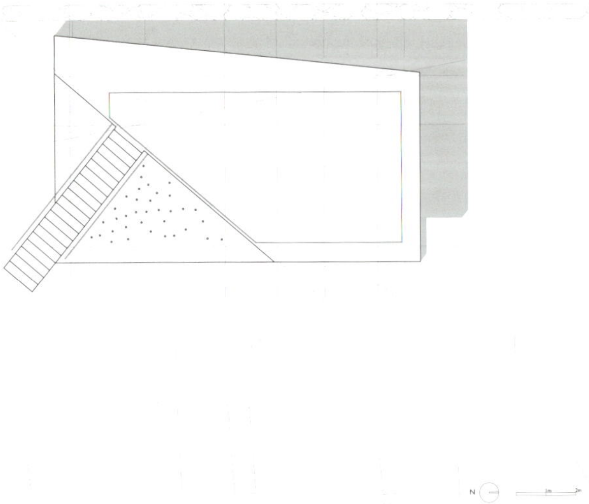

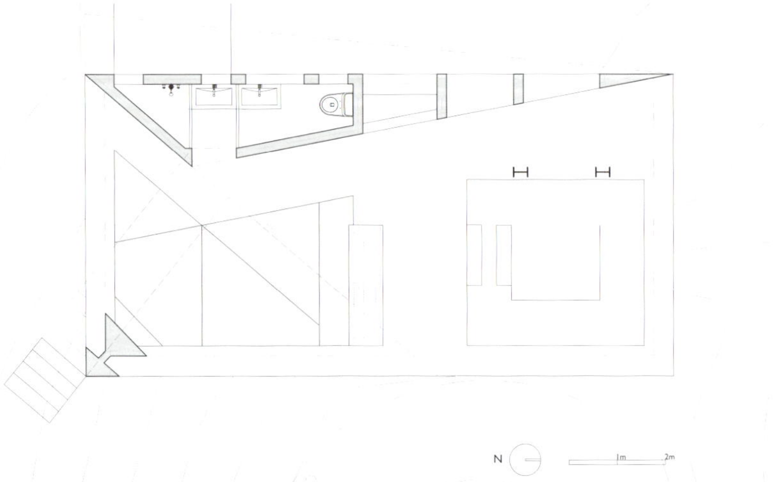

Plan/平面图

Casa Malinalco

Buscando La Aurora

- Architecture Design/建筑设计: Buscando La Aurora
- Project Architect/项目建筑师: Carlos Coronel, Hector de la Peña
- Location/地点: Malinalco, Estado de México
- Area/面积: site 1,000 m^2/buliding 400 m^2
- Completion/建成时间: 2008
- Photograph/摄影: BulAu

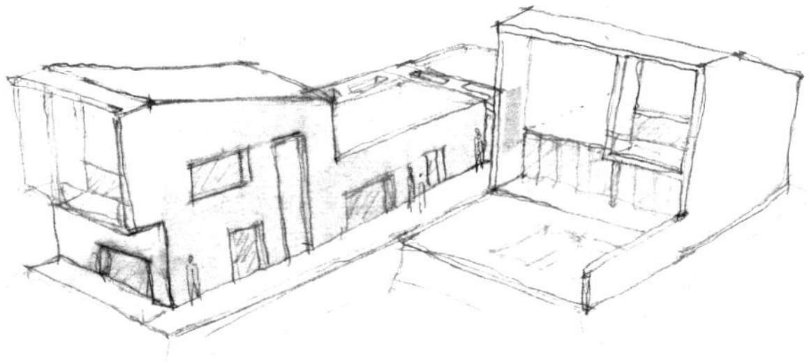

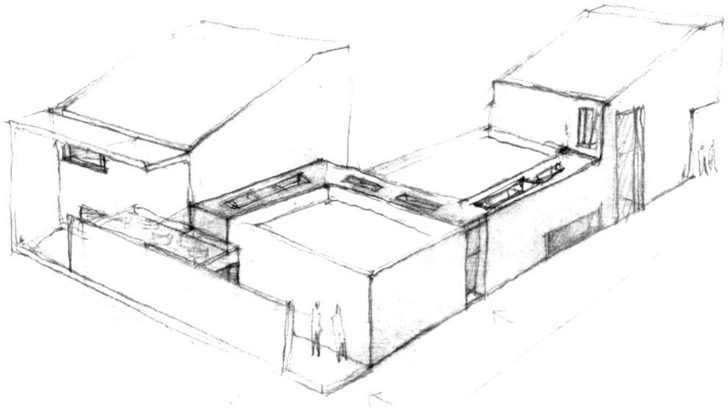

The project is located inside a well known golf club in Malinalco, a pre-Hispanic little town near Mexico city. The plot is a 41 x 23 meters oblique rectangle oriented east / west. It has two entrances, one through the street and one trough the golf course.

The house is an offset of the plot outline respecting the guidelines restrictions. The plan is divided in two volumes, one for public and one for private spaces, both articulated by a straight circulation, which runs from the entrance on the street side up to the golf course.

The dining and living rooms are in the smallest volume on the western side, the games and bar room is on the second floor, with a double height towards the dining area, where a wooden bridge floats over, connecting to the private volume, where its terrace becomes a deck with fabulous views to the mountains on the west. This deck terrace becomes the node between the rooms and the public area, since it has an entrance through the main room, which stands on the east side facing the golf course. The circulation becomes infinite, so the house can be experienced through an endless path. On the first plan of this volume three rooms and a tv room live to the central patio. The dining and living room area can be fully opened to integrate the inside to the outside.The façade walls are 30 cm. thick so they isolate the cold during the night and keep the heat gained during the day. Solar panels are used to heat the pool water and some bath services.

该项目位于墨西哥城附近的前西班牙小镇马利纳尔科的一个非常著名的高尔夫俱乐部内。这块地长41米，宽23米，是东西朝向的斜面矩形。该建筑有两个入口，一个是从街道上进入，另一个是从高尔夫球场进入。

这栋房子的设计弥补了地形的不足。设计师计划将这栋房子一分为二，即公共区域和私人区域，这两个区域仍然可以贯穿通行，马路边的入口可通向高尔夫球场的入口。

餐厅和客厅设在西侧最小的一部分空间里，游戏室和酒吧间设在二楼，其房间高度是餐厅区域的两倍。在那里设计了一个悬空的木桥，人们可以通过木桥走入私人区域，木桥还是一个可以眺望西面山峰很好的观景台。

这个观景台成为区分私人区域和公共区域的界限，从这里向东通过入口到主卧区，就可以观赏到高尔夫球场的全部景观。这样从东到西的室内设计营造出无限的空间感。人们沿着长长的通道，可以抵达房间的各个角落。

在第一部分的设计中，三间卧室和一个电视房是紧邻观景台的；餐厅和客厅区域也可以完全敞开，从而使室内和室外融为一体；建筑物正立面的墙有30厘米厚，因此它既能在夜晚抵御寒风，又能在白天吸收热量；太阳能电池板则用来加热泳池和一些洗浴设施用水。

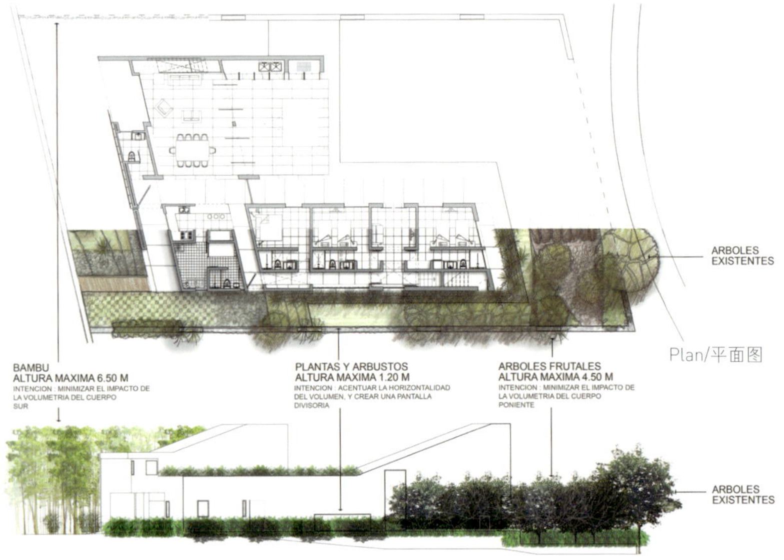

Plan/平面图

Facade: North/北立面图

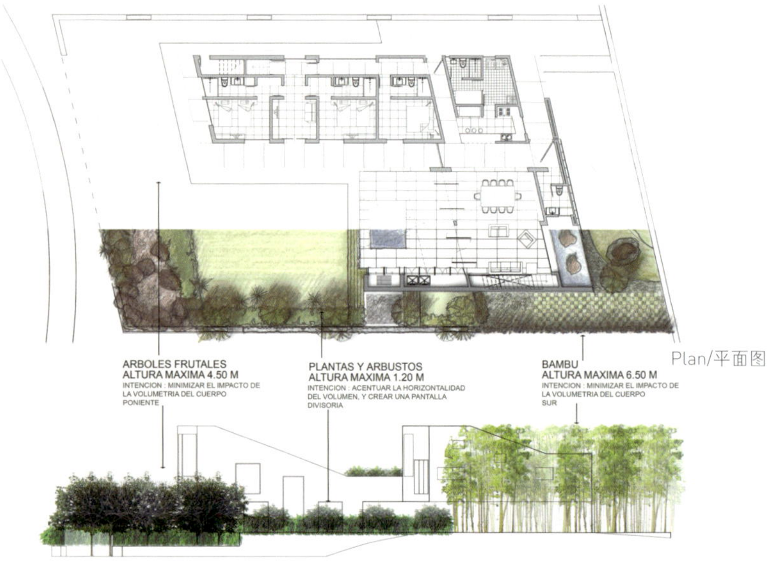

Plan/平面图

Facade: South/南立面图

Facade: West/西立面图

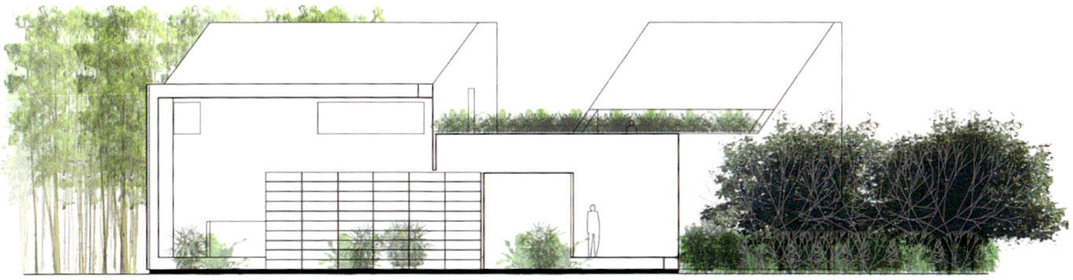

Facade: East/东立面图

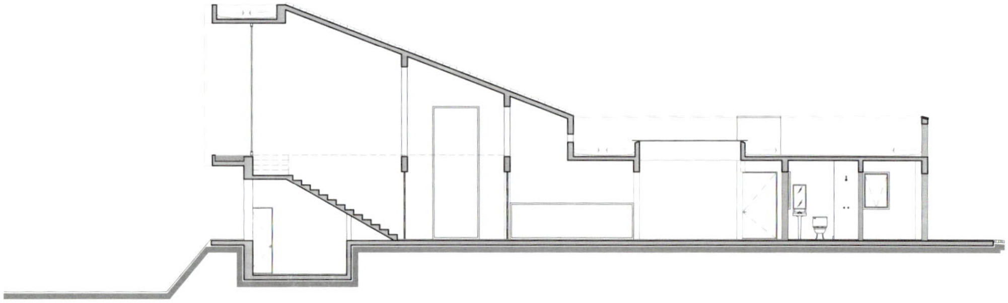

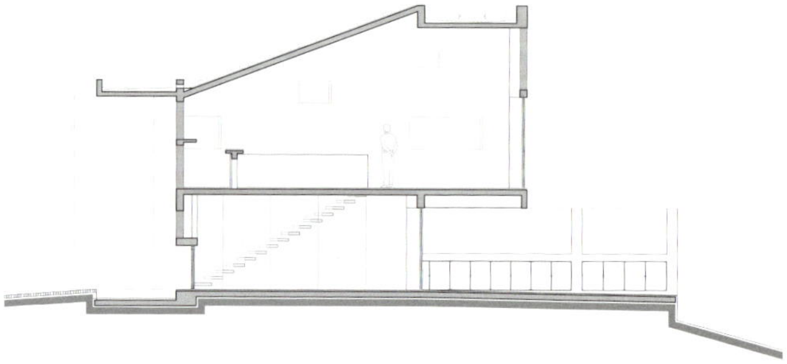

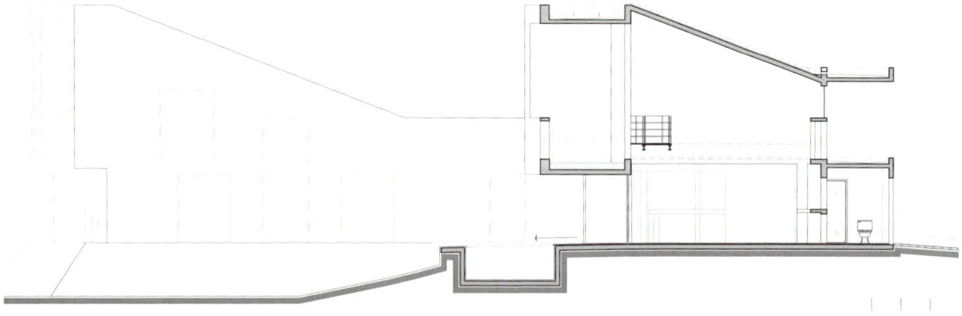

Section/剖面图

图书在版编目（CIP）数据

小空间 大智慧：小型绿色建筑设计实践 / ThinkArchit工作室 主编.
—武汉：华中科技大学出版社，2010.10
ISBN 978-7-5609-6527-7

I.①小… II.①T… III.①建筑设计-作品集-世界-现代 IV.①TU206

中国版本图书馆CIP数据核字（2010)第168947号

小空间 大智慧：小型绿色建筑设计实践　　ThinkArchit工作室 主编

出版发行：华中科技大学出版社（中国·武汉）
地　　址：武汉市武昌珞喻路1037号（邮编:430074）
出 版 人：阮海洪
策划编辑：杨　睿
责任编辑：龙　薇
责任监印：马　琳
装帧设计：张　羽
印　　刷：北京建宏印刷有限公司
开　　本：787 mm×1092 mm　1/12
印　　张：26
字　　数：300千字
版　　次：2010年10月第1版
印　　次：2010年10月第1次印刷
书　　号：ISBN 978-7-5609-6527-7 / TU·929
定　　价：318.00元（USD 78.00）

投稿热线：(010)64155588-8009　hzjztg@163.com
销售电话：(022)60266190,(022)60266199（兼传真）
　　　　　(010)64155566（兼传真）
网　　址：www.hustpas.com；www.hustp.com